Archimedes

New Studies in the History and Philosophy of Science and Technology

Volume 65

Archimedes has three fundamental goals: to further the integration of the histories of science and technology with one another; to investigate the technical, social and practical histories of specific developments in science and technology; and finally, where possible and desirable, to bring the histories of science and technology into closer contact with the philosophy of science.

The series is interested in receiving book proposals that treat the history of any of the sciences, ranging from biology through physics, all aspects of the history of technology, broadly construed, as well as historically-engaged philosophy of science or technology. Taken as a whole, Archimedes will be of interest to historians, philosophers, and scientists, as well as to those in business and industry who seek to understand how science and industry have come to be so strongly linked.

Submission / Instructions for Authors and Editors: The series editors aim to make a first decision within one month of submission. In case of a positive first decision the work will be provisionally contracted: the final decision about publication will depend upon the result of the anonymous peer-review of the complete manuscript. The series editors aim to have the work peer-reviewed within 3 months after submission of the complete manuscript.

The series editors discourage the submission of manuscripts that contain reprints of previously published material and of manuscripts that are below 150 printed pages (75,000 words). For inquiries and submission of proposals prospective authors can contact one of the editors:

Please find on the top right side of our webpage a link toour *Book Proposal Form.*

Sophia Kalantzakos
Editor

Critical Minerals, the Climate Crisis and the Tech Imperium

Springer

Editor
Sophia Kalantzakos
Environmental Studies
New York University Abu Dhabi
Abu Dhabi, United Arab Emirates

ISSN 1385-0180 ISSN 2215-0064 (electronic)
Archimedes
ISBN 978-3-031-25576-2 ISBN 978-3-031-25577-9 (eBook)
https://doi.org/10.1007/978-3-031-25577-9

This Springer imprint is published by the registered company Springer Nature Switzerland AG
The registered company address is: Gewerbestrasse 11, 6330 Cham, Switzerland

Foreword

This book covers an enormously important set of topics that provide foundational insights for the successful decarbonization and digitalization of the global economy. For governments, critical minerals have steadily risen on the political and policy priority list as demand has soared from cornerstone technologies such as electric car batteries. Additionally, as a result of some deep supply/demand imbalances and other market issues, prices have risen sharply for many minerals, and ongoing volatility looks likely. Opinions diverge as to whether the critical materials supply growth implied by various net zero scenarios can be met in the short and medium term. Thus emerges an important question: Will the lack of supply or the high prices of critical minerals slow down transitions to low-carbon technologies?

The field can be bewildering to outsiders. To begin with, there is no clear definition for critical minerals and materials. Various lists exist in several geographies, but they tend to change over time. The 30–100 unique minerals are part of myriad supply chains. The way "criticality" is measured typically relies only on supply side metrics and import dependency. The markets are largely small and opaque and offer neither good governance nor price discovery. Security of supply is important, but so too are the societal and environmental impacts as well as trade and market effects.

Adequate natural mineral resources exist. However, mining companies are often hesitant to invest in light of volatile markets, poor price signals, and highly uncertain policies. It is feared that current high prices may result in a flurry of new mine developments and a subsequent price collapse. This has happened before: a period of high rare earth prices 17 years ago resulted in 400 mine developments, but only one project made it to realization when prices collapsed again. Demand projections for 2030 and 2050 vary widely, thereby adding to the hesitancy.

The growth in demand for minerals and materials needed for the energy transition has already put a strain on supply. Mining and processing are the two key bottlenecks—and both deeply reliant on permitting and planning reforms. However, new capacity is not the only problem: the geographical concentration of where the mining and the processing is being conducted represent a key risk. China's dominance in processing, for instance, means economic or geopolitical shocks can disrupt markets. Discussions, however, tend to be fragmented and often focus on the minerals themselves. Rather, a holistic supply chain perspective is needed that includes advanced manufacturing and recycling.

Innovation is an important way to ease, or even radically change, demand for key materials. This includes battery chemistries, product design, processing technologies, recycling and circular economy concepts, new ways to incorporate sustainability in mining, as well as novel engagement with communities.

The markets for many of these minerals are thin and not transparent. As an example, volumes of lithium and REE are kt scale, compared to Mt-size copper, aluminum, or steel markets. The number of mines and processing is limited. Because of the small volumes, market growth to 2030 can be multiple times today's market volume. Thus, a few new mines can make a big difference, but the opacity, price discovery, and governance of such small markets will always present some difficulties.

Environmental and social concerns have to be addressed, as well. New mining operations are often controversial, and the development of a mine may take several years. The cancellation of the Rio Tinto Jadar project in Serbia in 2022 offers a salient example. The opposition of local communities continues to play a key role as it always has. A debate, moreover, is brewing around deep-sea mining and its environmental cost and benefits, notably for nickel, compared to mining in the biodiverse rainforests of Indonesia and the Philippines. There are also ongoing and significant governance issues in many of the critical minerals and products.

All critical materials are different, and there will be no "one size fits all" policy solutions. New forms of multinational cooperation are needed as well as partnerships with industry.

Finally, critical materials are a relatively new aspect for modeling and scenario development. This analytical work can help better understand the varied critical minerals supply chains and markets and inform investment and policy decisions. There is also a need for more comprehensive industrial ecology, materials flow analysis, life cycle analysis, trade flow analysis, and market analysis for critical materials.

The authors that contributed to the realization of this volume shed much needed interdisciplinary light on the range of challenges and opportunities as well as the complexities of the global green and digital transitions, accelerated by the urgency of the climate crisis and taking place in an era of hyper-competitive geopolitics. Importantly, their analyses help foster a conversation between the often siloed worlds of the academy, industry and finance, global institutions, and civil society.

Director of the Payne Institute for Public Policy and Professor at the Colorado School of Mines,
Golden, CO, USA
mbazilian@mines.edu
Morgan Bazilian

Lead for Hydrogen in the World Bank Energy Sector Management Assistance Program (ESMAP),
Washington, DC, USA
dgielen@worldbank.org
Dolf Gielen

Introduction

In July 2021, a cohort of scholars, experts from industry, finance, and global international institutions participated in a Summer Institute at Caltech and the Huntington called "Rich Rocks, the Climate Crisis and the Tech-imperium." It provided insights into the latest manifestation of resource competition. China's spectacular rise, the launch of its signature Belt and Road Initiative, the PRC's Made in China 2025 project (heavily focused on technological leadership and innovation), its growing assertiveness both economically and militarily, and its desire to make new norms and institutions have become areas of concern for the USA and other industrial nations, driving a wedge between them and accelerating power rivalries. As a result, the decarbonization and digitalization of the global economy, while no longer in question, is taking place in a climate of extremely fraught geopolitics that continue to deteriorate. Critical minerals are protagonists in this unprecedented economic and industrial transformation currently underway. Decarbonization, in fact, has become the main narrative driving COVID recovery. As of 2019, major economies have announced their goals to reach climate neutrality by mid-century. Political will combined with the rapidly declining production costs for clean technologies such as renewables and the electrification of transport are pushing the transition forward.

The assembled cohort examined different aspects of material criticality that have transformed these simple inputs for modern applications into materials of strategic and economic importance worthy of front-page news. Sessions were dedicated to the view from industry with a special focus on the nexus of government, industry, and finance. This is particularly significant given the growing push to reconstitute supply chains in ways that range from decoupling from China as much as possible to building resilience without rejecting interdependence. It also dived into an examination of the development and deployment of new technologies in light of the climate crisis and the next wave of digitalization. A session was devoted to the socioeconomic and environmental impacts of mining and processing of these important minerals which are mostly concentrated in the developing world. Moreover, participants discussed how the long-awaited green transition and the digitalization of the global economy could be more sustainable, inclusive, and equitable.

The Summer Institute led to the production of this volume which reflects an ongoing and fruitful collaboration among members of a diverse and rich cohort. It focuses on the nuts and bolts, the norms, values, economics, policy implications, environmental impacts, and geopolitics of the global green transition and digitalization drive. The volume is broken down into three distinct sections. The first tackles the geopolitics of decarbonization and digitalization especially in light of growing and acrimonious Sino-American competition. The second section brings together scholars and industry representatives in a discussion about critical minerals themselves and their supply chains. The material intensity of the transition and the possible bottlenecks are examined. Authors explore ongoing government responses and new pathways to secure uninterrupted access to the minerals and build resilient supply chains without severing ties of interdependence. In the third and final section, the mining of the materials themselves takes center stage. Authors discuss the environmental impacts, the role of ESG and smart mining strategies, and the pros and cons of deep-sea mining to diversify and augment the supply of minerals essential to decarbonization and digitalization.

In the opening chapter, Kalantzakos emphasizes that the decarbonization and digitalization of the global economy is taking place in an atmosphere of geopolitical upheaval. China's rise, its growing global impact, increased assertiveness, and more frequent heavy-handed autocratic policies launched under Xi Jinping's leadership have triggered an aggressive response by the USA, giving rise to a new cycle of bipolar competition. Alarmingly, in a spillover effect, the geopolitics of contention have polarized the transition to a decarbonized global economy and have made digitalization a battleground over who will control the tech imperium. Kalantzakos argues that this constitutes a dangerous development that undermines the formulation of an international coordinated response to the climate crisis which she views as the existential challenge for the global commons. Moreover, she argues, uninterrupted access to critical minerals (CRMs) that include rare earths (but also lithium and cobalt) are indispensable for both the decarbonization and digitalization of the global economy, but as a result of contentious and hypercompetitive geopolitics, a new race to control both the resources themselves and their extensive globalized supply chains is underway. Kalantzakos warns that the overuse of economic statecraft to settle a wide range of disputes is further fueling renationalizing narratives that seek to break ties of interdependence. In such a globally interconnected world economy, however, Kalantzakos argues that decoupling does not constitute a realistic option. Nor does it make the best strategic sense for either power. Her chapter recommends the lowering of the rhetoric of hyper-competition to allow the world the space for new possibilist thinking that recognizes interdependence and with a new geopolitics that includes the aspirations of both the developing and the developed world. In the end, she maintains that the response to the climate crisis must not become a starting pistol for a new scramble: for inputs, "geopolitically engineered" supply chains, the building up of tech and knowledge barriers, and new exclusions and inequities that engender securitized solutions to global challenges.

Pu's chapter focuses on how strategic competitions and political tensions are now shaping economic and technological ties between the USA and China. His chapter outlines the security concerns of both powers. Pu argues that China is deeply concerned that the USA is attempting to disrupt its domestic order and increasingly constrain its global reach. Moreover, according to Pu, the USA is principally worried about what it perceives as China's challenge to a US-led international order. Calls for commercial and technological decoupling from China have grown in Washington in the name of resilience and national security. China too is increasingly supporting a dual circulation strategy to build up its autonomy and offset US encirclement, emphasizing self-reliance and indigenous innovation. Pu's chapter provides argumentation for why a New Cold War narrative that is gaining momentum is both ill-advised and unrealistic. It would break the transnational production chain which would be too costly and unwelcome by other nations. Moreover, Pu posits that within a framework of competitive co-existence, both sides should build a recoupling scenario that he describes as partial recoupling. The author outlines the differences between partial recoupling from the notion of limited decoupling because the latter moves in the direction of reducing interaction and lessening interdependence. In Pu's view, this scenario would help preserve most normal economic ties between the two countries while carefully managing dependencies and risks. Given the growing rivalry and the short time frame for decarbonizing and digitalizing the global economy, Pu's analysis offers particularly helpful insights for policymakers who are concerned about securing uninterrupted global supply chains needed for the massive industrial transformation currently underway.

In her chapter, Vekasi argues that the high geographic concentration of critical minerals for the decarbonization and digitalization of the global economy has produced different policies and approaches to ensure uninterrupted access and build resiliency. Her chapter focuses on how best to secure supply chain resiliency for critical rare earth minerals. Critical minerals, including rare earths, face a rapidly increasing demand for the production of technology products and for the energy transition. China's rare earth industry has become more sophisticated and difficult to compete with in both mining and post-processing. Even though the demand for rare earths is projected to greatly rise, it unfortunately seems that the world is repeating the same debates and perhaps mistakes about supply risks. Her chapter outlines countries' efforts to reduce their dependence on Chinese rare earths, focusing particularly on the challenges and opportunities in diversifying supply and assessing possible paths forward. Vekasi describes the political economy of the rare earth market, focusing on the role of industrial and trade policy. She traces the policies that have resulted in the rare earth market becoming concentrated in China. She also examines the policies used in Japan and the USA to help suggest paths to greater independence, including diversification in production and capacity building, as well as talent retention in post-production processes. The chapter also provides an analysis of the political risks and challenges in geographic diversification along the supply chain. She finds that international cooperation in this sector is absolutely vital to meet future demand, a task that becomes more difficult given the strong anti-China rhetoric surrounding the PRC's role in the market. In her conclusion, Vekasi underscores that in the drive for increased resilience and supply, states must ready themselves to commit to investment in the full infrastructure of the industry, including education and training as well as sustainable new mining sites and processing projects.

In his chapter, Eggert raises the question about what kinds of public policies the USA and other market economies should have toward critical minerals. He rejects the false dilemma of the two "caricature like," mutually exclusive approaches to policy formulation on this issue. On the one hand, some are proponents of letting the market decide. They defend a sector-agnostic approach and argue for limiting government activities to establishing legal and regulatory frameworks for efficient and effective decision-making about private initiatives and activities. On the other hand, there are those who favor governments picking winning commercial sectors, companies, technologies, and products. Both these approaches ignore the most common and desirable middle ground. Eggert examines what constitutes this messy middle ground. He, moreover, provides a framework for and proposes seven principles to guide the development of public policy toward critical materials. Eggert underscores that market and industrial-policy solutions to supply-chain risks—rather than being mutually exclusive—are complementary. As an economist, he acknowledges his preference in anchoring public policy toward critical materials around market solutions but insists that they should also very selectively incorporate aspects of industrial policy.

In the next chapter, Karayannopoulos along with Tsianos provide readers a detailed and nuanced view from an industry perspective. Karayannopoulos, a veteran in the rare earth sector for several decades, witnessed how the global rare earth markets have ridden a dizzying roller coaster that has, at times, shaken the faith of industrial consumers in the reliability of supply chains for these critical materials. The chapter argues that currently rare earth production and associated supply chains are stronger and more diversified than at any other time in the last two decades. Still, the amount of materials will skyrocket, to meet rising demand driven by advanced technologies and global decarbonization. The authors argue that fundamental demand-driven industrial economics, coupled with a lack of coherent industrial policies in the West, led to the current supply chain map. It is the high growth of downstream industry participants and Original Equipment Manufacturing ("OEM") customers, either inside or in the proximity of the PRC, that have driven the rare earth industry's growth in China, not China's desire to dominate global rare earth markets as an end in itself. The chapter offers a different thesis to a dominant narrative that attributes the rare earth industry's current geographic concentration in Asia to an intersection of geopolitical forces and environmental policies. It claims, in fact, that supply chains are eminently rational and form efficiencies over time. In the case of rare earths that has meant that rare earth supply chains tend to develop in geographic proximity to their downstream customers and value-added manufacturing facilities, not necessarily near rare earth mining and processing facilities. Moreover, they conclude by examining industrial and policy strategies, that are being implemented with some success, to encourage the continued growth and resiliency of this important industry and detail ways to successfully establish resilient, diversified, and vertically integrated rare earth supply chains.

Berry's chapter makes an important point about the material intensity that the electrification of transport now requires. Both broad decarbonization and wider electric vehicle (EV) penetration will require more, not fewer, battery raw materials possibly raising the carbon intensity of industry, just as governments are coordinating a response to minimize carbon emissions. To hit even modest EV penetration rates (10% of new light-duty vehicle sales by 2025, for example) will require an even greater amount of raw material. Berry argues that innovation along the entire supply chain is a sine qua non in order to successfully decarbonize, and he calls for public-private sector collaboration to ensure that the resources necessary for further R&D are allocated. The chapter, moreover, looks at the challenges and opportunities facing the electric vehicle (EV) ecosystem as the industry embarks on a rapid growth phase. Berry emphasizes that there has been tremendous volatility in the pricing of battery metals since 2017, and this has contributed to much uncertainty for stakeholders along the lithium-ion battery supply chain. Lithium and cobalt in particular will be in even higher demand. As a result, technologies focused on sustainable battery production that include direct lithium extraction (DLE) or battery repurposing and recycling may offer viable solutions. Transparent pricing is key, but so is the identification of ways by which the governments can help the mining sector increase capacity sustainably and achieve decarbonization goals. Berry acknowledges that the unprecedented levels of extraction required for this transition raise questions about how to achieve the mining and refining of the materials while lessening the industry's carbon footprint.

In their chapter, Gauß, Gellermann, and Maurer warn that the ambitious government announced targets for reaching carbon neutrality by mid-century underestimate the extent and nature of the challenges involved in the access to vital materials for this massive transition. The "green deals" that will help achieve climate neutrality are dependent on quickly growing raw material demands, especially of primary metals, minerals, and advanced materials as well as biomass. All these resources will have to be extracted and produced, and come with significant environmental footprints. The authors argue that in the case of Europe, for instance, there is a growing danger of limiting policy formulation to a mere selection of technological choices in downstream manufacturing industry and energy sectors without considering the consequences in terms of costs, environmental footprints, and societal aspects along the entire life cycle of technologies and related value chains. More specifically, there is a real concern that Europe will become even more dependent on the import of advanced materials, overlooking the fact that it will also turn into an exporter of waste and CO_2 emissions to others. The chapter draws on Europe and Germany to illustrate the changing demand scenarios of supply chains needed to scale-up sustainable mobility and a bio- and hydrogen-based economy. Gauß, Gellermann, and Maurer argue that the European Commission has promoted sector-specific industrial alliances to foster public-private partnerships in accelerating innovation and sustainable industrial growth. Examples are, the formation of the

European Battery Alliance launched in 2017, the European Raw Materials Alliance launched in 2020, and the Chips Act of 2022.

In her chapter, Klinger addresses the environmental impacts of rare earth mining and processing. She raises the question of whether an upfront investment in mitigating social and environmental hazards that may result from rare earth mining and processing is more or less expensive than cleaning up after the fact. While seemingly straightforward, the question itself raises a host of other issues, particularly around how costs and responsibilities for them are defined and delimited. Klinger uses qualitative, historical, environmental, and epidemiological research from a range of sectors to outline the basis of comparison between *ex ante* and *ex post* pollution controls in the rare earth sector. Following an introduction to the problem, the chapter reviews the contemporary social and environmental impacts of rare earth mining, mainly in China, and considers the political economic frameworks that limit effective mitigation of social and environmental harm as well as supply chain diversification. Klinger then assesses and provides a preliminary scope for evaluating the social and environmental costs of mining within and beyond the site of extraction, and on this basis compares the possible cost parameters for *ex post* vs. *ex ante* mitigation of social and environmental harms. In conclusion, her chapter discusses the need for further research efforts and open publication of data and results, in order to accurately assess the costs of rare earth mining and processing, to calculate the difference between *ex ante* and *ex post* pollution controls, and to guide effective policy formulation and implementation.

In his chapter, Ali argues that even though extractive industries have had immense social and environmental impacts, they remain the foundation of modern society. However, as the climate crisis worsens necessitating global economic decarbonization and digitalization, a number of minerals have become absolutely vital for the production of green technologies. Ali argues, however, that the opportunity to extract minerals on land is diminishing due to community opposition to new mines. He offers examples of several countries such as El Salvador and the Philippines that have moratoria on new mining projects. Even in the USA, opposition to mining remains high, and while the Trump presidency was receptive to extractive industries, a large metal mine in Alaska was rejected due to concerns about pollution impacts. In this context, the author posits that oceanic mineral resources are now being looked at as an important option since there are no direct social impacts in the deep sea. The oceans remain a vast but vulnerable frontier for scientific and industrial inquiry. The Anthropocene, moreover, continues to bring new challenges and opportunities for ocean conservation. Ali acknowledges that ecological impacts are largely unknown and remain a point of concern for conservationists. This has been a momentous time for the International Seabed Authority (ISA), which was formed over two decades ago as part of the United Nations Law of the Sea Convention, because it is scheduled to release a set of environmental standards and guidelines related to the implementation of its mining code in 2023. In this chapter, therefore, Ali provides a review of some of the key issues that can help highlight points of contention for more informed decision-making. Moreover, he considers the ways in which international environmental governance of the extractive industries is developing to address conflicts and how more effective decisions can be made in consideration of social and environmental tradeoffs.

In his chapter, Pell points to how the accelerated efforts to decarbonize the economy because of the climate emergency have spotlighted lithium's importance as an input in the global economic transformation. As a result, there is growing scrutiny, not only over the way that lithium is being mined and the massive amounts of material that will be needed moving forward, but also across the entire supply chain, from mine to market. Pell's chapter focuses on Latin America's "Lithium Triangle" which accounts for over 60% of global lithium reserves where extraction is concentrated in brine deposits and uses water-intensive processes that affect both environmentally fragile salt flats and adjoining biodiverse wetlands—areas that are also linked, both economically and culturally, to Indigenous populations. Pell argues that the management of lithium mining challenges will be changing due to the recent Escazu Agreement (2018) and emerging ESG disclosure and diligence requirements, particularly from the European Union (EU), whose standards are the most developed. His chapter examines ways by which the Escazu Agreement and EU ESG standards provide a paradigm for how companies can be encouraged to pay greater attention to long-term environmental and social issues, and the role that partnerships among industry, government focal points, and civil society organizations (CSOs) could play in maximizing ESG value strategies.

We would like to express our deepest gratitude to Jed Buchwald and Dan Lewis for their support of our cohort of the Summer Institute at Caltech and The Huntington and the production of this volume. Warm thanks to Steve Hindle and Jean-Laurent Rosenthal and to the wonderful Caltech staff that facilitated our meetings.

Sophia Kalantzakos

Environmental Studies
New York University Abu Dhabi
Abu Dhabi, United Arab Emirates
sophia.kalantzakos@nyu.edu

Contents

About the Editor

Sophia Kalantzakos is a Global Distinguished Professor in Environmental Studies and Public Policy at New York University Abu Dhabi. Her research centers on the geopolitics of critical minerals, the transition to a net zero future, and the fourth industrial revolution. Her work, in particular, examines how resource competition in an era of fraught geopolitics has tilted the balance toward more securitized assessments of global interdependence. Moreover, she examines China's global aspirations manifested in the Belt and Road Initiative, Europe's reckoning with a seismic push against both its normative and economic power, and the US's re-evaluation of its leadership role in the global order. Kalantzakos' work examines how current epistemic systems will need to give way to new modes of thinking. As nation-states are turning inward in response to demands for de-globalization even while future challenges remain intensely global, her work advances the construction of dynamic, inclusive, and action-oriented responses. Her publications include *China and the Geopolitics of Rare Earths* (Oxford University Press, 2018; rev.2021) and The EU, US, and China Tackling Climate Change: Policies and Alliances for the Anthropocene (Routledge, 2017). She was a Rachel Carson (RCC) Fellow at LMU Munich in 2015 and 2018 and has served as a member of its Academic Advisory Board. She is currently the president of the RCC Society of Fellows. Recently, she has been a Fung Global Fellow at the Princeton Institute for International and Regional Studies (PIIRS) and Senior Fellow in the Research Institute for the History of Science and Technology at Caltech and The Huntington.

Part I
Geopolitics of Decarbonization and Digitalization

Chapter 1
Between Rocks and Hard Places: Geopolitics of Net Zero Futures and the Tech Imperium

Sophia Kalantzakos

Abstract The geopolitics of contention centering around growing US-China hyper-competition has polarized the transition to a decarbonized global economy and has made digitalization a battleground over control of the tech imperium. This dangerous development undermines the realization of a coordinated international response to the climate crisis. It has also led to the securitization of access to critical minerals and their extensive globalized supply chains, jeopardizing the manufacturing of technological end products at a time when an already narrow window for constructing new infrastructure and deploying green technology is rapidly closing. The overuse of economic statecraft to settle a wide range of disputes (exacerbated by the Russian invasion of Ukraine) has, moreover, further fueled renationalizing narratives that seek to break ties of interdependence. This chapter provides insight into the latest manifestation of resource competition. It recommends the lowering of the rhetoric of hyper-competition so that the response to the climate crisis does not become a starting pistol for a new scramble: for inputs, "geopolitically engineered" supply chains and the building up of tech and knowledge barriers that produce new exclusions and inequities engendering securitized solutions to global challenges.

Keywords Interdependence · Reshoring · Rare earths · European Battery Alliance · Net zero · Critical minerals · EVs

Two seismic economic and technological shifts are taking place in a climate of extremely fraught geopolitics. The global economy is set to urgently decarbonize because of the climate crisis, but also simultaneously digitalize thus ushering in the fourth industrial revolution. The transition constitutes a pivotal moment for global economic transformation. Unfortunately, though importantly, both these monumental shifts are happening in an atmosphere of geopolitical upheaval. First, China's growing global impact, increased assertiveness, and more frequent heavy-handed

S. Kalantzakos (✉)
Environmental Studies, New York University Abu Dhabi, Abu Dhabi, UAE
e-mail: sophia.kalantzakos@nyu.edu

S. Kalantzakos (ed.), *Critical Minerals, the Climate Crisis and the Tech Imperium*, Archimedes 65, https://doi.org/10.1007/978-3-031-25577-9_1

autocratic policies launched under Xi Jinping's leadership have triggered an aggressive response by the United States, giving rise to a new cycle of bipolar competition which is negatively impacting the workings of the world order. Second, the Russian invasion of Ukraine in early 2022 resulted in the use of unprecedented levels of economic statecraft and triggered extraordinary energy and commodity shocks. Prices of coal, oil, and gas soared to levels that surpassed previous energy crises. By the end of 2021, even before the invasion of Ukraine began in earnest, prices of gas were ten times those of 2020 in Europe and Asia, while they tripled in the US. The continued supply chain disruptions because of COVID outbreaks and growing transportation costs, moreover, also slowed the growth of renewable energy projects worldwide, as this particular industry was designed to be global from the outset.

While one hundred and thirty five countries have adopted some form of a net zero target, governments all over the globe scrambled to keep the lights on, formulating policies to help diversify fossil fuel imports, especially if they were overly reliant on Putin's Russia as the EU has been. Fossil fuel production domestically was also ramped up wherever possible, and energy was subsidized to shield consumers who were particularly hard hit by these developments. China, for instance, announced an increase of coal production, the US experienced a boom of fracking and drilling projects, while the EU explored and introduced caps on the prices of fossil fuel imports and other policies to alleviate the economic strain to citizens and industry (REN21 2022). Making matters even worse, the summer of 2022 saw unprecedented forest fires, floods, droughts, and extreme weather conditions. It fueled confusion and urgency on how to proceed with the green energy transition as the addiction to fossil fuels had clearly resulted in states' prolonged energy vulnerability. Finally, the Russian invasion, China's reluctance to condemn Putin's actions, and Washington's more assertive support of Taiwan, drove an even bigger wedge between the US and the PRC to the point where the two powers halted, albeit temporarily, collaboration on climate leadership in August 2022 (Xu 2022).

Nonetheless, even under these chaotic conditions both the US and the EU pushed through new bills (Repower EU and the Inflation Reduction Act) meant to support the green transition, to build energy autonomy, reduce emissions, create jobs, and tackle inflationary pressures on the economy (European Commission 2022a; The White House 2022a). Political will combined with the rapidly declining production costs for clean technologies such as renewables and the electrification of transport continue to push the transition forward. At the same time, calls to diversify, reshore, and renationalize supply chains which had already started when the US threw down the gauntlet to China, continued to grow. Interdependence once lauded as an asset became a liability, a vulnerability that could be weaponized by rivals.

An explosive geopolitical landscape that centers around Sino-American hypercompetition thus formed and continues to worsen. Importantly, the transition to a net zero future has turned into a nationalist competition for normative, economic, technological, and military pre-eminence. Moreover, it appears that the response to the climate crisis has become a starting pistol for a new scramble: for inputs, "geopolitically engineered" supply chains, the building up of tech and knowledge barriers, that produce new exclusions and inequities engendering securitized solutions to global challenges (Kalantzakos 2020a).

Alarmingly, the geopolitics of contention has not only polarized the transition to a green global economy, but has also made digitalization a battleground over who will control the tech-imperium (Piore 2022). As a result, these transitions are no longer confined to the normal parameters of trade competition. They constitute an open and heated rivalry over influence, economic power, supply chains, information, norms, values, and governance. This represents a dangerous development because the rhetoric of division (framed as a dilemma between autocracy vs democracy) (Biden 2021a) is eclipsing the existential challenge for the global commons which is to produce an international coordinated response and blueprint to offset the worst impacts of the climate crisis and the disruption of the earth's systems.

Critical minerals are necessary protagonists in this unprecedented economic and industrial transition if the global community truly aims to keep temperatures from rising beyond 1.5 °C by cutting greenhouse emissions by 50% by 2030. This deadline is both fixed and pressing. There is, therefore, an existential crisis to solve, a terribly tense geopolitical atmosphere now entangled in normative dilemmas of autocracy vs democracy, and a short window of time left to realize this industrial transformation while also maintaining growth, jobs, and profits.

Moreover, digitalization accelerated significantly because of the pandemic, and forced many sectors to go virtual requiring the deployment of artificial intelligence (AI) and 5G networks. This ushered in the fourth industrial revolution earlier than initially anticipated. The deployment of 5G, for instance, will expand mobile networks to support a diversity of both services and devices. They will redefine a range of industries retail, education, transportation, and entertainment. AI, moreover, is being rapidly developed promising to accelerate innovation, efficiency, and sustainability in businesses and organizations. These complex systems that underpin the next industrial revolution are voracious consumers of critical minerals (Greenfield and Graedel 2013).

China's competitors and rivals have awoken to the realization that the PRC has again been better prepared. It has strategically invested in resources and globalized supply chains for key inputs and applications that are vital for the global economic transformation. Its signature Belt and Road Initiative, moreover, that re-imagines the ancient silk road, also builds maritime connections to directly bridge and streamline geographically, economically, territorially, and perhaps even politically, Eurasia and Africa while also looping in South America (Kalantzakos 2020b). The BRI thus forms a novel network of land, maritime, and digital connectivity. Importantly, industrial innovation and production are no longer the exclusive purview of the OECD economies, especially the US, European Union member states, and Japan. Compounding this new reality is the fact that the long era of stable resource competition is rapidly ending. Historically, empires locked in their economic supply chains and managed competition. And in the long post-1945 cycle of decolonization, the United States, as the global economic hegemon, was the power that defended and ensured that the rules and norms of world trade were enforced (Kalantzakos 2020a). China's growing consolidation of its commercial and military interests as well as its expanding network of partnerships have changed the game. Governments in the developing world where a number of critical minerals are concentrated have welcomed China's one-stop-shop financing schemes, increased engagement, and narrative of "win-win partnerships" thereby wresting away from traditional actors both influence and sway (Shambaugh 2020).

The global economic recovery is, thus, inextricably linked to this complicated and politically highly charged transition. Uninterrupted access to critical minerals (CRMs) that include rare earths (but also lithium and cobalt) are indispensable for both the decarbonization and digitalization of the global economy. Regrettably, the contentious and hypercompetitive geopolitics has unleashed a new race to control both the resources themselves and their extensive globalized supply chains.

The level of interdependence that characterizes the globalized resource supply chains demand that they remain unobstructed to ensure the uninterrupted production of these applications. Critical minerals are all highly geographically concentrated and vulnerable to disruption. Moreover, the amounts needed will skyrocket. A low-carbon future will be very mineral intensive because clean energy technologies need more materials than fossil-fuel-based electricity generation technologies. According to a 2020 World Bank report the production of lithium and cobalt may increase by 500% by 2050 to meet clean energy demand alone (Hund et al. 2020). This means that the number of new mining projects will be unprecedented if we are to secure the levels of materials required (Arboleda 2020).

The numbers speak for themselves. We will be further increasing the use of renewables in the global energy mix to break off our dependence on fossil fuels and power our economies. In 2021 as the economy began to recover, a record 314.5 GW of new renewable power capacity was added. In practical terms it was enough to power every household in Brazil. Globally installed renewable power capacity, moreover, reached 3146 GW. Solar PV and wind accounted for nearly 90% of all new renewable power additions. In 2021, the share of renewable energy in electricity generation reached 28.3% (REN21 2022).

In terms of transportation, the electrification of vehicles will help significantly reduce CO_2 emissions. Transport according to the IEA in 2021 accounted for 37% of CO_2 emissions from end-use sectors (International Energy Agency 2022a). And while in 2021 EV sales globally reached nearly 10%, major global car manufacturers are moving rapidly into EV production. In 2020 sales of global plug-in vehicles reached 3.2 million and in 2021, 6.6 million (International Energy Agency 2022b). Even with these significant increases in sales, however, EVs account for only about 1% of the global fleet which means that the pace of production and sales will only increase (Bradsher 2021; Gorner and Paoli 2021; Powell 2021; REN21 2022). China and Europe are leading with the most new EV and plug-in registrations followed by the United States. The market for fast electric recharging stations will also be growing in light of electric vehicle sales (Strategic Market Research 2022).

Lithium and cobalt are indispensable inputs for the electrification of transport. Demand for lithium tripled between 2015 and 2018 in order to keep up with the production of lithium-ion batteries used in hybrid and EVs, but also in electric tools, grid storage applications, and portable electronic devices. In 2021 production reached 100,000 tons (Jaskula 2022). By 2022, demand was expected to exceed supply a trend that would continue at least to 2025 and even 2030 (Berry 2022). Producers have therefore ramped up production. Lithium prices also jumped significantly; spot prices for lithium carbonate in China climbed 170% in 2021, to RMB142,000 ($22,000) a ton, their highest since April 2018. In June 2022, it was already at $60,000 per ton. Prices of spodumene, a source of lithium mainly mined in Australia, climbed 144% in 2022, to $990 a ton and 5, 250 per ton in June 2022 (Argus 2022; SMM 2022).

By the same token, cobalt is a key ingredient in batteries, smartphones, laptops, and electric cars. In 2020, The Democratic Republic of Congo (DRC) continued to supply 70% of the world's mine production (U.S. Geological Survey 2021). The DRC, one of the poorest countries of Africa, also boasts by far the largest reserves in the world estimated at three times that of Australia (U.S. Geological Survey 2021). The price for cobalt also went up in 2021 because of demand. More specifically, according to a 2021 IRENA report, cobalt recovered from a low of 30 USD/kg in 2019 to surpass 70 USD/kg in late 2021 but then in 2022, it reached $80/kg (Gielen 2021; Argus 2022).

One would think given how globalized economies have become, that the market will take care of supply when there is growing demand. Prices will fluctuate, new mines may open, industry will find substitutes, and perhaps recycling could help respond to shortages. These are reasonable responses to supply and demand dynamics; that is if we weren't also faced with renationalizing narratives that seek to break ties of interdependence and if geopolitics hadn't led to the overuse of economic statecraft to settle a wide range of disputes (Lew and Nephew 2018; Norris 2016; Atlantic Council 2021).

In fact, one such instance in which economic statecraft was perceived as having been used by China to settle an incident with Japan flagged the importance of rare earths for high tech, defence, and green applications. Rare earths are 17 elements of the periodic table that are part of the family of lanthanides. They are costly to mine and often toxic to separate if environmental regulations are not stringent. But they are absolutely essential ingredients in lighting, magnets, glass, electronics, defense systems, wind turbines, and a slew of other technologies. Yet for all their importance rare earths were virtually unknown. They became notorious overnight, after a trawler incident in 2010 involving a Chinese vessel and the Japanese coast guard around the contested (by China) Senkaku (Diaoyu) islands quickly turned into a geopolitical dispute provoking an "unofficial" and short-lived embargo of rare earths to Japan (Bradsher 2010a, b). It was a wake-up call for major industrial actors that for the first time recognized an important vulnerability that could be exploited if China decided to flex its muscles (Vekasi 2019a).

The inordinate attention paid to China's control of rare earths was not merely a political reaction to a geopolitical play. It was a response to the shocking realization that there were few options outside China's control where other countries could source rare earths. This had not always been the case. The United States was initially the leader in the production and trade of rare earths and in finding ever advanced technological uses for them. The discovery of rare earths in Mountain

Pass California in 1949 initially gave the US the upper hand. Then came China's discovery of exceptionally rich rare earth deposits in Bayan Obo in inner Mongolia and with it an important appreciation of their growing strategic importance. This led to the development of a comprehensive strategy that included increased investment in their production and metallurgy as well as in supply chains for new important rare earth applications (Vekasi 2019b). By the 1990s the PRC had effectively strangled the competition and dominated the industry (Hurst 2011; Klinger 2017). In 2010, China controlled 97% of worldwide production; nine years later in 2019, 85–93% of production, not a major dent, all things considered. Following the rare earths crisis, the two largest alternative non-Chinese mines, Molycorp in Mountain Pass California went bankrupt in 2015 and Lynas the biggest mine in Australia was salvaged at the last minute by Japanese investment. Even today, China continues to dominate rare earth production holding at around 80–85%. They also consume about 2/3 of global rare earth output. Molycorp restructured and went public again in November of 2020 as MP Materials, attracting both investment and DOD contracts in response to calls for diversification. Still its feed, at the time of writing, is shipped to China for processing in facilities of one of its key investors -- Chinese Shenghe Resources. They are not alone. China is also the default destination for junior rare earth miners' output of raw materials and intermediate products. In terms of key supply chains, China's market share in NdFeB (neodymium iron boron) magnets, for instance, is close to 90%.

How could developed nations allow a near monopoly of such a strategic resource to be in the hands of a powerful competitor? If anything, the 2010 rare earth crisis provided a salient case study of how resource competition would look in the future. It also spotlighted the fact that most of China's competitors were largely depending on the market to deliver solutions, whereas China had developed its rare earths sector openly yoking private and public partnerships to meet its goals.

Dependency on China for rare earths was never fully tackled because prices normalized quickly enough and quotas limiting exports were abolished after the case was brought to the (WTO 2015). Companies returned to China to purchase the rare earths they needed because it made economic sense in the context of the need to maintain their competitive share prices and protect their bottom line. Mining for rare earths elsewhere and rebuilding global supply chains was, in the end, too cumbersome, too expensive, and necessitated long lead times. Moreover, governments lost interest and turned to other more "pressing matters" because global power relations at that juncture while increasingly competitive were not as polemical nor had persistent impediments to global trade reared their head. Many analysts characterized the entire crisis as a blip and moved on (Johnston 2013).

But then the trade wars began. China's rise was now projected as being a major threat to the existing power structure. The world was at the dawn of another major economic transformation and OECD countries wondered what it would mean if they lost their competitive edge to China. This sobering realization led the most advanced economies a decade later to again scramble for ways to build up their stocks, find alternatives sources of REEs, and diversify supply chains. Only now, the list of critical minerals has expanded to include among others lithium and cobalt.

Lithium resources are highly concentrated in South America – in Argentina, Chile, and Bolivia -- that form what is known as the Lithium Triangle (cf. Pell Chap. 10 below), even though at present Australia is the largest worldwide producer. Chinese companies have pursued mine investments in both South America and Australia to ensure that they maintain a dominant position in downstream industries as well as an overall command of the supply chain. Similarly, with respect to cobalt, China again stands out for developing a strategy that secures its position of influence over the industry and ensures that it remains the dominant actor in cobalt mining in the Congo (Kalantzakos 2021).

China has over decades been actively consolidating its relationships in the developing world and secured a dominant position in the extraction industry. China's competitors that had previously dominated industrial innovation and production, are consequently feeling exceedingly vulnerable. Major industrial nations have again updated their critical minerals lists, are attempting to build resilience against possible disruptions, and are seeking to re-constitute supply chains closer to home. In the United States decarbonization politics turned into urgent calls to desinicize supply chains in order to thwart China's ability to control the CRM space.

Under Trump, breaking interdependence and supply networks reliant on China was priority number one. On September 30, 2020, Trump signed an Executive Order that addressed the "Threat to the Domestic Supply Chain from Reliance on Critical Minerals from Foreign Adversaries," calling it a national emergency (Executive Order 13953 2020). Three years earlier, in December 2017, Trump had also signed Executive Order 13817: A Federal Strategy to Ensure Secure and Reliable Supplies of Critical Minerals (Executive Order 13817 2017). It tasked the Federal government with making policy "to reduce the Nation's vulnerability to disruptions in the supply of critical minerals." The Trump administration announced plans to collaborate with Australia to finance and develop CRM projects in 2019 (Reuters 2019) and the State Department launched the Energy Resource Governance Initiative (ERGI) as a critical material initiative (U.S. Department of State 2019).

Anti-China sentiment grew further under the Biden Administration. On February 24, 2021, Biden signed an executive order asking for a "report identifying risks in the supply chain for critical minerals and other identified strategic materials, including rare earth elements (as determined by the Secretary of Defense), and policy recommendations to address these risks" (The White House 2021a).

In Alaska, during the first high level meeting between the US and China that took place in March 2021, the animosity was palpable (Nikkei Asia 2021). Moreover, having made it his goal to urgently decarbonize the US economy and lead in the production of new green technologies, just like his predecessor, Biden continued to underwrite efforts to build US resiliency prioritizing the desinicization of supply chains (Rogers and Plumber 2021).

In his March 2021 press conference, Biden threw down the gauntlet to the Chinese. "The future," he said, "lies in who can in fact own the future as it relates to technology, quantum computing, a whole range of things, including the medical fields. China is out-investing us by a long shot because their plan is to own that future. China has an overall goal, and I don't criticize them for the goal, but they have an overall goal to become the leading country in the world, the wealthiest country in the world, and the most powerful country in the world. That's not going to happen on my watch" (Biden 2021b). In June 2021, the findings of the "critical minerals supply chain review" were made public. The report identified America's areas of weakness along the supply chains of critical minerals and recommended the preparation of a strategy to secure the strategic and critical minerals supply chain in a way that "expands sustainable production and processing, while at the same time working with allies and partners to secure the global supply chain" (The White House 2021b). In December 2021, moreover, Biden signed the Uyghur Forced Labor Prevention Act into law that had bipartisan support. The law bans imports from the Uyghur-majority Xinjiang autonomous region in northwestern China and imposes sanctions on foreign individuals responsible for forced labor in the region (Sonmez 2021). Before 2021 expired, Biden held a Democracy Summit in which he underscored that the US was "committed to working with all those who share those values to shape the rules of the road," and added that the United States would stand by those "who give their people the ability to breathe free and not seek to suffocate their people with an iron hand" (Taipei Times 2021; Hilman 2021). He pledged US$424 million to support media freedom, fair elections, and anti-corruption campaigns. In January 2022, "Restoring Essential Energy and Security Holdings Onshore for Rare Earths Act of 2022" was introduced in the US Senate, that would force defense contractors to stop purchasing Chinese rare earths by 2026. If passed, the bill would use the Pentagon in order to create a permanent stockpile of strategic minerals (Scheyder 2022). Standing up to China became a policy supported across the aisles.

In such a globally interconnected world economy, however, decoupling from an economic giant like China does not constitute a realistic option. Nor does it make the best strategic sense. While wary of China's growing economic footprint and more open peddling of Xi Jinping's model of effective governance for the twenty-first century, Asian countries - many allies of the United States - are resisting the division of the world in two rival camps. While welcoming a more globally engaged Washington, the domestic political situation in the US raises skepticism as to what might happen if either Trump or someone that espouses Trump's "America First" policy comes back into office. While the midterm elections of 2022 did not result in a red wave, domestic politics in the US remain highly volatile allowing some to question whether the US will still be a democracy by 2030 (Dixon 2021).

Case in point: In November 2020, China and the member states of ASEAN (Association of Southeast Asian Nations) plus Australia, Japan, South Korea and New Zealand signed the Regional Economic Comprehensive Partnership (RCEP), (The Ministry of Foreign Affairs of Peoples Republic of China 2020) because Trump had left the US led Transpacific Partnership (TPP) as soon as he took office (Baker 2017). It was a moment that the PRC government used to drive the point that "ideas such as decoupling with China are nonsense and … nations who pursue such are likely to end up on the outside of the world's economic gravity" (Lee 2020).

China is also signaling that the Belt and Road Initiative will remain at the center of its global vision. As Chinese Foreign Minister Wang said during his 2021 Africa tour, "China will work with all parties to deepen the synergy between the BRI and the development strategies of other countries, keeping in mind their needs for COVID-19 response and economic recovery" (New China TV 2021). New belts and roads are being added with digital, medical, tourism and education among the top priorities (Li and Shapiro 2020).

By the same token, the European Union, the United States' closest ally, militarily, politically, and economically also signed a major trade deal with Beijing in December of 2020 after the Biden election (Blenkinsop 2020). The EU views China as a systemic rival but prefers the deepening of multilateralism and the setting of common standards with its partners. In Europe's view it can stay true to its values and principles and pursue "likeminded partnering" to form the most relevant multilayer group possible when dealing with any specific issue.

Admittedly, relations between the EU and China became more difficult following a barrage of tit for tat measures over events in Hong Kong and the treatment of the Uyghurs that put the trade deal in deep freeze (Ridgwell 2021). Relations between Europe and the PRC further worsened as a result of China's "ambiguous" position framed as "non-interference" regarding the Russian invasion of Ukraine. This development provoked a further urgency to diversify supply chains and triggered more vocal skepticism about the enduring nature of Sino-European cooperation. It also led to a renewed European strategic rapprochement with the United States. Still, the EU did not pull the plug on its partnership with the PRC. European Commission President Ursula von der Leyen, European Council President Charles Michel, and High Representative Josep Borrell expressed European concerns to China's leadership over the possibility of Beijing offering more practical support to the Putin regime. In contrast to the Biden-Xi summit on that same issue, however, the EU did not point a finger at the PRC but focused on the centrality of EU-China trade relations, areas of overlapping strategic interest, and extensive opportunities to deepen collaboration. In their talks with then Prime Minister Li Keqiang and subsequently with President Xi Jinping, EU leadership offered a transparent and detailed explanation of the enormous significance that the outbreak of a war on the European Continent posed for EU member states (Liboreiro 2022). It simultaneously kept the agenda more open to include other interrelated matters such as cooperation on global food security, climate change, and the fight against COVID (European Council 2022).

It is instructive to underscore Europe's distinct approach to both the decarbonization and digitalization of its economy since it differs both substantively and rhetorically from that of the United States. Europe speaks about growing supply chain resilience but has also strategically chosen - unlike the United States - to actively attract global investment, build new networks of interdependence, and strengthen existing ones without alienating the PRC. In 2017, Europe put its words into practice. It formed the European Battery Alliance to address the need for efficient batteries that were essential for transport, power, and industrial applications. Seven hundred industrial and innovation actors from mining to recycling have come together under the EBA to help build a strong and competitive European battery industry (European Battery Alliance 2020). This initiative demonstrated that it is possible for the EU to strengthen its position as a producer of applications for the new low carbon economy without severing networks of interdependence. It is not, therefore, surprising that Chinese and other Asian companies are already investing in Europe because the opportunities for collaboration are highly attractive from a business standpoint (Preisinger and Bryan 2018). In January 2021, for instance, the European Commission approved €2.9bn in state aid funding to support research and innovation in the battery value chain in coming years, as part of a new European battery innovation project. The funds would be provided from 12 member states including Austria, Belgium, Croatia, Finland, France, Germany, Greece, Italy, Poland, Slovakia, Spain, and Sweden. The commission's expectation, moreover, is that the project will attract a further €9bn in private investments. In other words, numerous member states have now a real stake in the deployment of this transition (European Commission 2021a).

The EU has followed a similar logic in launching the European Raw Materials Alliance in the Fall of 2020. ERMA which is the largest consortium in the raw materials sector worldwide is designed to support a multi-sourcing strategy of REEs, ensure resilient and transparent supply chains, and increase European industrial competitiveness (European Raw Materials Alliance 2021). ERMA's first goal is to focus on the most pressing needs: increasing the EU resilience in the rare earth magnet and motor value chain.

The reason for this is that 95% of electric vehicles use traction motors containing rare earth magnets. Over 100,000 tons of rare earth permanent magnets are used each year in renewable energy, machine tools, robotics, loudspeakers, water pumps, mobility, and ICT. Moreover, 16,000 tons of rare earth permanent magnets are exported from China to Europe each year, representing approximately 98% of the EU market (Gauß et al. 2021). These figures spotlight Europe's potential vulnerability and existing dependence on the PRC, but a low-carbon future will in any case be extremely mineral intensive. ERMA clearly prioritizes securing uninterrupted access to rare earth elements that are indispensable to magnet production, particularly neodymium, praseodymium, dysprosium, and terbium. These four rare earths constitute only 25% of the total rare earths production volume but represent 80–90% of the total rare earths market value. Moreover, clean energy technologies need more materials than fossil-fuel-based electricity generation technologies.

In line with the EU Commission mandate, the EU's objective is to secure access to sustainably produced rare earth magnets at competitive costs from primary and recycled sources; to make Europe a global leader in rare earth metal, alloys, and magnet production; and to sustain and expand Europe's global leadership in electric motor and generator design. In the case of rare earths, for example, the only industrial-scale commercially-operating separation facility in Europe – based in Estonia – is 70% dependent on Russian feedstock. The 30% balance comes from a unique supply chain arrangement between the US-based Uranium company Energy Fuels and Neo Performance Materials, utilizing by-product chemistry and economics to make the business case work. Over the next 4–5 years, as Neo Performance Materials develops the Sarfartoq mining deposit in Greenland, European rare earth magnet supply chains will be able to diversify even further and establish optionality

(Neo Performance Materials 2021). To achieve even further diversification, in November 2022, the Estonian government awarded Neo 18.7 million euros from the EU Just Transition Fund to launch a magnet plant in Estonia (Neo Performance Materials 2022).

Moreover, in June of 2021, the EU and Canada jointly announced the Canada-EU Strategic Partnership on Raw Materials that focuses on the integration of Canada-EU raw materials value chains, science, technology and innovation collaboration as well as efforts to advance world-class environmental, social and governance criteria, and standards. These efforts are meant to encourage private and public investments between the two parties and advance resilient raw material and downstream value chains. The EU and Canada intend their partnership to also strengthen alignments in strategic downstream industry, particularly batteries, clean technologies, and other advanced manufacturing sectors (European Commission 2021b).

With respect to the growth of AI, the EU is cognizant of the fact that it may find itself sandwiched between two combating technological giants, the US and China. Both are gearing for full-fledged confrontation. America's attack on Huawei was just the beginning. In October of 2020, the Trump administration placed new restrictions on exports to Semiconductor Manufacturing International Corporation, China's most advanced maker of computer chips. In the summer of 2020, tik tok went from being a mere platform of dance videos produced by teenagers to a geopolitical threat, when the US government (Swanson et al. 2020), at the time, threatened to ban it implying that it served as a surveillance mechanism in the hands of China (Rogers and Kang 2021).

While President Biden reversed Trump's order on tik tok, he called for a broader review of a number of foreign-controlled applications that could pose a security risk to Americans and their data. Washington has clearly launched an orchestrated and multifaceted pushback against China. In fact, in the summer of 2021, Biden expanded Trump's order that "banned U.S. investment in Chinese companies that support China's military to include those selling surveillance technology, calling the entities a threat to U.S. interests and values" (Whalen and Nakashima 2021; The White House 2021c). In April 2021, the US Senate, moreover, introduced the *Strategic Competition Act*, a bipartisan supported bill to mobilize all United States strategic, economic, and diplomatic tools for an Indo-Pacific strategy that enables the U.S. government to compete effectively with the People's Republic of China and the challenges it poses to US national and economic security (U.S. Senate 2021). In June, the Senate, again with bipartisan approval, passed a bill that "aims to boost U.S. semiconductor production and the development of artificial intelligence and other technology in the face of growing international competition, most notably from China" (NPR 2021). The Chips and Science Act, as it became known, was signed into law by Biden in August 2022 allocating $52 billion (The White House 2022b).

Given the tensions brewing between the two competitors, and eager to set standards and rules, already in 2018 the European Union drafted a *Coordinated Action Plan on Artificial Intelligence* that supports ethical, secure, and cutting-edge AI made in Europe (European Commission 2018; Hauser 2021). It also passed a Chips

Act in August 2022 dedicating an overall sum of 43 billion euros (drawing from other existing programs) to ensure the EU's semiconductors ecosystem, increase its resilience, as well as ensure supply and reduce external dependencies (European Commission 2022b). Both the US and the EU are reliant on production of chips especially for next gen nano chips of the 2 nanometer range on Asian companies in Korea and Taiwan specifically. While the US is strong in the design technology, the EU's sole advantage lies in lithography and semiconductor fabrication as ASML holds a near monopoly in the EUV machinery market (Castellano 2020).

In western media, the power rivalry is amplified by countless stories that stoke fears that China will cut off the world from critical minerals to demonstrate geopolitical strength and overtake the US in the digital race. After the Russian invasion, the securitization of the discourse surrounding both the minerals and their supply chains, but also the digital transition, have become further augmented resulting in more urgent calls for a robust plan of action under the specter of impending confrontation. There is shockingly little discussion on how fraught geopolitics undermine multilateral efforts to address the worsening climate crisis threatening the global commons. This indicates the extent of how magical thinking has made it somehow possible for major competitors to believe that they can disagree on everything "significant," but simultaneously find room for cooperation on the issue of the climate crisis, as if this is some peripheral, soft issue and not the greatest challenge of our time.

Additionally, the stories in the press fail to problematize the socio-economic impacts of the race for minerals on the developing world and overlook the unprecedented ecosystem degradation on land and sea that frenzied extraction will cause. Nor is it clear that these "new" supply chains will make trade more environmentally or socially sustainable. Even more egregious is the silence surrounding the need for knowledge sharing and building the production capacity for these technologies with and in the developing world. They turn a blind eye, to the developed world's promise to create green jobs and bolster their economies in order for them to genuinely leapfrog into the next economic transformation.

Moreover, even the geopolitics of critical minerals is not as linear as portrayed, but far more complex. Nuance, however, hasn't proven as effective as the polarizing narrative which triggers "urgent," though often haphazard and ineffective, government responses. Reporting has systematically ignored China's perspective which would have been instructive and enlightening. To begin with, the PRC has not been hiding its ambitious plans to decarbonize and digitalize its economy and to become a leading producer of new technologies. In the regime's view, it is hard to rival strong existing industries. There is more opportunity to become a leader in new technologies that are not yet saturated. China is intent on leading not only in EV production but also in AI technology.

According to the International Data Corp, China's AI investment is expected to reach US$26.69 billion in 2026. This accounts for about 8.9% of global investment, that ranks China as second in the world (IDC 2022). According to Wang Yu, research fellow at the College of Intelligence and Computing at Tianjin University, "AI will first play a key role in helping industries to achieve intelligent production and enable green and low-carbon development to occur faster." (Cheng 2021) With AI, cities will also be more intelligent. By excelling in the deployment of AI the hope is that China can peak carbon emissions by 2030 and achieve carbon neutrality by 2060, allowing it to reach that goal faster than anticipated. China's declared intention to be an AI leader is met with escalating push back by the United States. More and more frequently, press reports and government announcements address China's attempts to control Big Data. The more hawkish have demanded that the US government launch a broad review of Chinese internet, telecom, and tech companies operating in the country, and restrict activities—and the ability to access American data for the security American economic, military and commercial interests (Piore 2022; Pottinger and Feith 2021).

China's Made in China 2025 industrial plans require a higher supply of rare earths for domestic production. In fact, since 2018, the PRC has become an importer of rare earths, particularly of the more valuable ones such as praseodymium and neodymium used in magnets (Li and Jia 2019). Following the 2010 rare earth crisis, moreover, China sought to consolidate and streamline the industry in order to reduce environmental pollution, but also to curb smuggling of REEs outside the PRC. This was a crucial government objective because domestic stability relies on a social contract that values continued economic growth and effective pollution reduction. It is likely that the 2019 Xi Jinping visit to a rare earths mining site with his chief US-China trade negotiator in the midst of the trade war with the US may have been a not so subtle attempt to rattle the US administration resulting in headlines that suggested that the PRC might be ready to "weaponize" the elements (Rogers et al. 2019). Still, China has not pulled that trigger (since 2010) even while it continues to protect and strengthen its overall strategic position in the rare earths sector. New alarms were again raised in 2021, when China announced further standardization of the rare-earth industry to promote its high-quality development. In January 2022, it also made public the creation of a new "megafirm" (Qian and Brooke 2022) China Rare Earth Group Co. Ltd. which will have control of 60–70% of Chinese rare earth production, translating into 30–40% of global supply. Nonetheless, it is long-term planning and China's ability to produce rare earths at a fraction of the cost compared to the few mines outside the PRC that have made it very difficult to break this advantage.

There is no doubt that critical minerals and their supply chains are now at the center of geopolitical strife with businesses caught in the crossfire. While tensions are often stoked by junior miners that see an opportunity for government support in the prevailing climate, companies by and large are eager to see governments lower the political rhetoric and look at the bigger picture. For companies, both accessible markets especially large ones like the US, the EU, and China and stable politics are essential to their planning and growth. Key companies involved in the decarbonization and digitalization of the global economy have recognized that, independent of what causes them, long stretches of supply chain disruption and bottlenecks hurt their operations and their bottom line. The pandemic already illustrated the ways that economic life can be upended. Businesses are, therefore, in favor of building more resilient supply chains and necessary infrastructures closer to home. In many instances, they have now recognized and accepted that there will be a cost to pay for this premium. Car companies, for instance, that are quickly shifting to the production of electrical vehicles, understand that alternative sources of processed rare earths will be considerably more expensive, and they are factoring this in their design to diversify their supply options. Still, while economic competition is expected, a narrow realist frame of pitting countries' national interests against the other creates disruption and uncertainty which also adversely impacts economic performance. Case in point, the Russian invasion of Ukraine caused further supply chain disruptions, wildly fluctuating energy spikes, exorbitant increases of commodity prices, shortages, inflation, and an unprecedented use of economic statecraft in order to force Russia to stop the violence. With respect to technology, in the early days of the war, shortages in the global supply for semiconductors, for instance, continued to worsen impacting the auto sector and consumer electronics. Neon, for example, is produced in Ukraine and constitutes a vital input for the manufacturing of advanced chips. Disruptions in nickel production further increased EV battery costs. As a result, diversification and resilience against exogenous shocks became more urgent while interdependence went from being considered an asset to being viewed as a growing liability since it can be weaponized.

Still, if the US, the EU, and other major industrial nations are serious about diversifying their supply chains, industry will need to be willing to pay a premium. Governments will have to subsidize private initiatives in ways that will encourage the capital investments necessary to produce magnets, in America and/or Europe. The previous rare earths crisis indicated that companies and governments were not willing to finance diversification. Today, things may be changing because of the race to decarbonize and to massively scale up the production of EVs and renewable applications. ESG criteria may also incentivize companies to underwrite the cost of alternative, but costlier supply chains. Time is of the essence since 2030 is a pressing deadline by which the world can stop temperatures from rising more than 1.5 °C. It remains to be seen which companies step up to the plate, and how governments can aid the building of necessary infrastructures for midstream processing stages.

More troubling still is that the basis on which the strategy of managing a risen China has failed to account for and appreciate the complexity of Beijing's calculations and preparation for the existential challenges of the twenty-first century. First, the PRC has increasingly begun to interpret Western actions as attempts at encirclement and tightening containment. Second, Beijing is acutely aware of the levels of havoc that the climate crisis may cause the State's stability and coherence. Afterall, the domestic implications resulting from the climate emergency motivated the government to take widespread action to reduce pollution (air and water) and rapidly decarbonize China's economy. Though skeptics of China's sincere commitment to its climate goals abound, it is a mistake to overlook the thinking that permeates Beijing's policymaking. The climate crisis is the existential threat to the PRC's overall stability and prospects, and in response China has been fortifying its defenses to both protect itself internally even while expanding its outreach, through its regional and global networks. In other words, Beijing is busy creating protective walls, a fortress for the Anthropocene. Some infrastructures (such as the build-up of its maritime defense capabilities, the creation of a string of pearls, the territorialization of atolls in the South China Sea, the damming of its rivers) are visible while some others are more fluid (the great firewall of China, the digital belt and road, the cultural, educational, and medical belt and road).

Fortress China is principally a project designed for a world that will soon operate in chaotic conditions in which its competitors' decisions most likely will be a series of crisis management responses to new situations as the climate crisis deepens. As a government whose principal concerns are political, that is to ensure domestic stability and the continuation of one-party rule, Beijing is playing a multi-tiered game, simultaneously planning and preparing for multiple challenges, both expected and unexpected (Xi 2022). China's plans to secure and maximize its advantage face pushback and suspicion especially since the rivalry has taken on a highly normative tone centered around a face-off between autocracy and democracy. But there are lessons to be garnered by looking at China's plans through a climate lens that offer a better explanation of the PRC's endgame. It would also allow for ways to think of geopolitical priorities, strategies, and responses for the Anthropocene.

Will these developments threaten the global resolve to decarbonize? Will digitalization further deepen great power rivalry? Will governments (i.e. tax payers) be called upon to subsidize and support the transition so that it continues to make "economic sense" for industry? These are key questions that will affect the resolve and commitment to tackle the climate crisis and will trigger contentious debates on budget allocation.

First, a good paradigm to examine and emulate is that of the EU. The European Union has done an impressive job in its decarbonization strategy and roll-out despite geopolitical upheaval. Internally, Europe has committed to no net emissions of greenhouse gases by 2050; to decoupling economic growth from resource use, and to leaving no person and no place behind in this transition. Its Green Deal aspires to prepare all sectors of the EU economy in meeting this challenge and maps out the path to fulfill the pledge to reduce emissions in a way that is fair, cost effective, and competitive.

Europe has demonstrated that climate constitutes its signature foreign policy ambition. It has steadfastly kept climate negotiations alive, even though – ironically- the US has repeatedly seized the spotlight to announce Sino-American global leadership on climate. The Union has pursued a policy of new global collaborations. In the summer of 2021, the EU and Canada, for instance, signed off on a strategic partnership to produce both materials and applications needed for decarbonization. The EU approach offers a number of instructive insights for the US, despite the United States being divided on the climate agenda.

In Washington, critical minerals had, for a long while, been narrowly linked to national defense, even though the EV revolution and the production of green technologies are the true consumers of these inputs. It is only recently that there has been a shift in rhetoric as the Biden administration is more assertively leaning toward the urgency of the transportation and energy transition following the signing of the Inflation Reduction Act of 2022. The administration is thus also leveraging the Defense Production Act (DPA) to solve both national security and industrial policy economics for critical minerals needed for EVs. It is now recognized that the market size for defense applications alone is not large enough to justify the existence of domestic supply chains for many critical minerals. Moreover, given that the US aspires to secure a major economic share of the global decarbonization and digitalization market, decoupling narratives may need to be reframed as the pursuit of effective collaboration and sharing competencies with close US partners. Still while the DPA recognized Canada as a "domestic" supplier and there is a proposal to grant Australia and the UK the same status so that their projects for strategic and critical materials are eligible for funding, that is not the case for the EU which is not at present being considered eligible for DPA Title 3 applications (U.S. Department of Defense 2022; Peters et al. 2022).

More importantly, even though the US is pursuing decoupling strategies, China is expanding its portfolio of investments in the Americas. In 2022, for instance, JL MAG RARE-EARTH CO., LTD announced that it would set up a new company to invest in the construction of the "Project for Comprehensive Utilization of Waste Magnetic Steel" in Mexico with total planned investment of approximately US$100 million (JL MAG RARE-EARTH CO., LTD 2022). Upon completion 5000 tons of waste magnetic steel per year would be comprehensively utilized, and the capacity of producing 3000 tons of high-end magnetic material products per year would be developed. There is noticeable push back by US allies, moreover, who are not willing to sacrifice the China market. In September of 2022, for instance, Japanese auto producer Honda, announced that it will set up an EV battery purchasing venture with Chinese car manufacturer Dongfeng Motors and Guangzhou Automobile to optimize its battery supply chain (Argus Media 2022).

Second, though numerous initiatives to build resilience are currently under discussion, they have largely kept major powers pre-occupied with siloed strategies. Securitizing initiatives have been playing out, for instance, through a re-invigorated QUAD, the Five Eyes intelligence sharing Alliance, and The National Technology and Industrial Base (NTIB) group that centers its attention on issues of national security and dual-use research and development. Critical minerals and their supply chains have been on the agenda of the G7, the G20, and the OECD. These fora have reinforced the securitization and fragmentation of the decarbonization drive and fuel geopolitical upheaval. What is perhaps needed is the identification of another institution better suited to spearhead a multilateral response. To this end, some suggest that the International Energy Agency (IEA) take the lead, given its established track record. It is already mobilizing its convening power and dedicating resources to building a strategy vis-a-vis critical minerals. Others propose IRENA, an intergovernmental organization with a vibrant membership, located in Abu Dhabi which is dedicated to the energy transition (Ali et al. 2022). Whether IRENA fits the bill warrants further exploration, but it would strengthen non-securitized multilateral collaboration and help ensure that the green transition is more just and inclusive.

Third, the debate on how to ensure access of critical minerals should also include strategies on how to rebuild relations (not military alliances, but civil relations) with the developing world. Europe never left the developing world, especially Africa, but the US did pull back after the end of the Cold War. It may now be the opportune moment to design a more comprehensive diplomatic and investment approach that avoids resuscitating a core-periphery division. The Biden administration attempted this more novel approach through the *Build Back Better Initiative*, but the project has stalled indefinitely. In contrast, the EU announced *Global Gateway* and remains determined to see it succeed.

Finally, major economic powers will have to come to terms with the fact that there can be no leadership without climate leadership. To lead one needs some framework, a narrative that summarizes its goals and intentions. At the moment, there are three competing narratives vying to become "universal." China's *ecological civilization*, Europe's *living well within the limits of our planet*, and an emerging American one on *environmental justice*. In their own way, each of them proposes a vision for the future and helps think beyond the black and white normative dilemmas of democracy versus autocracy.

In the end, the response to the climate crisis must not become an excuse for a new scramble. Climate leadership and geopolitical confrontation cannot and must not be thought of as compatible. Pretending to do both only serves to perpetuate paradigms of global power imbalances and socio-economic injustice that have led us to the brink of planetary catastrophe.[1]

References

Ali, S.H., S. Kalantzakos, R. Eggert, R. Gauß, C. Karayannopoulos, J. Klinger, X. Pu, K. Vekasi, and R.K. Perrons. 2022. Closing the infrastructure gap for Decarbonization: The case for an integrated mineral supply agreement. *Environmental Science and Technology* 56: 15280–15289. https://doi.org/10.1021/acs.est.2c05413.

Arboleda, Martin. 2020. *Planetary mine: Territories of extraction under late capitalism*. Brooklyn: Verso.

Argus. 2022. Argus Metals International: Global non-ferrous market prices, news and analysis. Issue 22-107, 8 June.

Argus Media. 2022. Japan's Honda plans EV battery venture in China. *Argus Media*, 8 September. https://www.argusmedia.com/en/news/2289110-japans-honda-to-build-battery-ev-plant-in-china. Accessed 10 Sept 2022.

Atlantic Council. 2021. Reality check #2: Economic sanctions should not always be the go-to foreign policy tool, 22 February. https://www.atlanticcouncil.org/content-series/reality-check/reality-check-2-economic-sanctions-should-not-always-be-the-go-to-foreign-policy-tool/. Accessed 2 Sept 2022.

Baker, Peter. (2017). Trump abandons trans-Pacific partnership, Obama's Signature Trade Deal. *The The New York Times*, 23 January. https://www.nytimes.com/2017/01/23/us/politics/tpp-trump-trade-nafta.html. Accessed 2 Sept 2022.

[1] This chapter originated as the 2021 annual William and Myrtle Harris Distinguished Lecture in Science and Civilization at Caltech. I benefited greatly from the public discussion that ensued. I also wish to thank Jeremy Adelman for illuminating exchanges on critical minerals during my time as a Fung Global Fellow at the Princeton Institute of International and Regional Studies in 2019–2020. In addition, I am grateful to NYU Abu Dhabi for generously supporting my research.

Berry, Chris. 2022. Interview with expert, September 2022, cf. chapter 6 in this volume.
Biden, Joe. 2021a. Biden's speech to congress: Full transcript. *The New York Times*, 29 April. https://www.nytimes.com/2021/04/29/us/politics/joe-biden-speech-transcript.html. Accessed 11 June 2021.
———. 2021b. President Joe Biden first white house press conference transcript. *The White House*, 25 March. https://www.rev.com/blog/transcripts/president-joe-biden-first-white-house-press-conference-transcript-march-25. Accessed 10 Sept 2022.
Blenkinsop, Philip. 2020. EU agrees investment deal with China to rebalance ties. *Reuters*, 30 December. https://www.reuters.com/article/us-eu-china-trade/eu-agrees-investment-deal-with-china-to-rebalance-ties-idUSKBN2941AP. Accessed 10 July 2022.
Bradsher, Keith. 2010a. Amid tension, China blocks vital exports to Japan. *The New York Times*, 22 September. https://www.nytimes.com/2010/09/23/business/global/23rare.html. Accessed 2 Sept 2022.
———. 2010b. Rare earths stand is asked of G-20. *The New York Times*, 5 November. http://www.nytimes.com/2010/11/05/business/global/05rarechina.html. Accessed 2 Sept 2022.
———. 2021. As cars go electric, China builds a big lead in factories. *The New York Times*, 4 May. https://www.nytimes.com/2021/05/04/business/china-electric-cars.html. Accessed 10 Sept 2022.
Castellano, Robert 2020. ASML: Not just a monopoly in EUV lithography. *Seeking Alpha* 15 June. https://seekingalpha.com/article/4354007-asml-not-just-monopoly-in-euv-lithography. Accessed 2 Sept 2022.
Cheng, Yu 2021. AI to usher in brand new age. *China Daily*, 5 November. https://global.chinadaily.com.cn/a/202105/21/WS60a70d23a31024ad0bac073a.html. Accessed 1 Nov 2022.
Dixon, Homer. 2021. Opinion: The American polity is cracked, and Might Collapse. Canada Must Prepare. *The Globe and Mail*, 31 December. https://www.theglobeandmail.com/opinion/article-the-american-polity-is-cracked-and-might-collapse-canada-must-prepare/?utm_medium=Referrer:+Social+Network+/+Media&utm_campaign=Shared+Web+Article+Links. Accessed 19 June 2022.
European Battery Alliance. 2020. Building a European Battery Industry. *European Battery Alliance*. https://www.eba250.com/. Accessed 15 Nov 2020.
European Commission. 2018. Coordinated Plan on Artificial Intelligence. *European Commission*, 7 December. https://digital-strategy.ec.europa.eu/en/library/coordinated-plan-artificial-intelligence. Accessed 19 June 2022.
———. 2021a. State Aid: Commission Approves €2.9 Billion Public Support by Twelve Member States for a Second Pan-European Research and Innovation Project along the Entire Battery Value Chain. *European Commission*, 26 January. https://ec.europa.eu/commission/presscorner/detail/en/IP_21_226. Accessed 19 June 2022.
———. 2021b. EU and Canada set up a strategic partnership on raw Materials. *European Commission*, 21 June. https://single-market-economy.ec.europa.eu/news/eu-and-canada-set-strategic-partnership-raw-materials-2021-06-21_en. Accessed 18 Sept 2022.
———. 2022a. REPowerEU. 18 May. https://ec.europa.eu/commission/presscorner/detail/en/IP_22_3131. Accessed 10 June 2022.
———. 2022b. European chips act factsheet. *European Commission*, 8 February. https://digital-strategy.ec.europa.eu/en/library/european-chips-act-factsheet. Accessed 10 June 2022.
European Council. 2022. EU-China summit: Restoring peace and stability in Ukraine Is a Shared Responsibility. *European Council*, 1 April. https://www.consilium.europa.eu/en/press/press-releases/2022/04/01/eu-china-summit-restoring-peace-and-stability-in-ukraine-is-a-shared-responsibility/. Accessed 3 Apr 2022.
European Raw Materials Alliance (ERMA). Homepage. *ERMA*. https://erma.eu/. Accessed 23 Feb 2021.
Executive Order 13817. 2017. A federal strategy to ensure secure and reliable supplies of critical minerals. *Federal Register*, 20 December. https://www.federalregister.gov/documents/2017/12/26/2017-27899/a-federal-strategy-to-ensure-secure-and-reliable-supplies-of-critical-minerals. Accessed 1 Sept 2022.

Executive Order 13953. 2020. Addressing the threat to the domestic supply chain from reliance on critical minerals from foreign adversaries and supporting the domestic mining and processing industries. *Federal Register*, 30 September. https://www.federalregister.gov/documents/2020/10/05/2020-22064/addressing-the-threat-to-the-domestic-supply-chain-from-reliance-on-critical-minerals-from-foreign. Accessed 12 Sept 2022.

Gauß, Roland, et al. 2021. Rare earth magnets and motors: A European call for action. *ERMA*. https://eitrawmaterials.eu/wp-content/uploads/2021/09/ERMA-Action-Plan-2021-A-European-Call-for-Action.pdf. Accessed 1 Sept 2022.

Gielen, Dolf. 2021. Critical materials for the energy transition. *IRENA*, May. https://www.irena.org/Technical-Papers/Critical-Materials-For-The-Energy-Transition. Accessed 12 Sept 2022.

Gorner, Marine, and Leonardo Paoli. 2021. How global electric car sales defied Covid-19 in 2020. *IEA*, 28 January. https://www.iea.org/commentaries/how-global-electric-car-sales-defied-covid-19-in-2020. Accessed 10 June 2021.

Greenfield, Aaron, and T.E. Graedel. 2013. The omnivorous diet of modern technology. *Resources, Conservation and Recycling* 74 (May): 1–7. https://doi.org/10.1016/j.resconrec.2013.02.010.

Hauser, Hermann. 2021. The struggle for technology sovereignty in Europe. *Project Syndicate*, 12 March. https://www.project-syndicate.org/commentary/europe-technology-sovereignty-imperative-by-hermann-hauser-2021-03. Accessed 1 Sept 2022.

Hilman, Jennifer. 2021. The transatlantic relationship could make or break Biden's summit for democracy. *Council on Foreign Relations*, 1 December. https://www.cfr.org/in-brief/transatlantic-relationship-could-make-or-break-bidens-summit-democracy. Accessed 15 Jan 2022.

Hund, Kirsten, et al. 2020. *Minerals for climate action: The mineral intensity of the clean energy transition*. International Bank for Reconstruction and Development/The World Bank. http://pubdocs.worldbank.org/en/961711588875536384/Minerals-for-Climate-Action-The-Mineral-Intensity-of-the-Clean-Energy-Transition.pdf

Hurst, Cindy. 2011. *China's rare earth elements industry: What can the west learn?* vol. 2011. 20 August. https://fmso.leavenworth.army.mil/documents/rareearth.pdf

International Data Corporation. 2022. China's artificial intelligence market will exceed US$26.7 Billion by 2026, according to IDC. *IDC*, 3 October. https://www.idc.com/getdoc.jsp?containerId=prAP49740122. Accessed 20 Oct 2022.

International Energy Agency. 2022a. Transport: Improving the sustainability of passenger and freight transport, *IEA*. https://www.iea.org/topics/transport. Accessed 2 Dec 2022.

———. 2022b. Global EV Outlook 2022, *IEA*. https://www.iea.org/data-and-statistics/data-product/global-ev-outlook-2022. Accessed 2 Dec 2022.

Jaskula, Brian. 2022. *Lithium*. U.S. Geological Survey, January. https://pubs.usgs.gov/periodicals/mcs2022/mcs2022-lithium.pdf. Accessed 1 Sept 2022.

JL MAG RARE-EARTH CO., LTD. 2022. *Proposed investment in the construction of the 'project for comprehensive utilization of waste magnetic steel'*. JL MAG RARE-EARTH CO., LTD, 9 September.

Johnston, Alastair Iain. 2013. How new and assertive is China's new assertiveness? *International Security* 37 (4): 7–48.

Kalantzakos, Sophia. 2020a. Critical minerals and the new geopolitics. *Project Syndicate*, 2 October. https://www.project-syndicate.org/commentary/china-critical-minerals-new-geopolitics-by-sophia-kalantzakos-2020-10?barrier=accesspaylog. Accessed 1 Sept 2022.

———. 2020b. The race for critical minerals in an era of geopolitical realignments. *The International Spectator* 55 (3 (July)): 1–16. https://doi.org/10.1080/03932729.2020.1786926.

———. (2018, 2021) *China and the geopolitics of rare earths*. New York: Oxford University Press.

Klinger, Julie Michelle. 2017. *Rare earth Frontiers: From terrestrial subsoils to lunar landscapes*. Ithaca: Cornell University Press.

Lee, Yen Nee. 2020. China Is Not an 'expansionist Empire,' State Media Says after Signing of Mega Trade Deal. *CNBC*, 16 November. https://www.cnbc.com/2020/11/16/chinese-state-media-on-the-signing-of-worlds-largest-trade-deal-rcep.html. Accessed 29 Aug 2022.

Lew, Jacob J., and Richard Nephew. 2018. The use and misuse of economic statecraft. *Foreign Affairs*, 15 October. https://www.foreignaffairs.com/articles/world/2018-10-15/use-and-misuse-economic-statecraft. Accessed 10 Sept 2022.

Li, Liuxi, and Denise Jia. 2019. China Became Net Importer of Rare Earths in 2018. *Caixin Global*, 16 March. https://www.caixinglobal.com/2019-03-16/china-became-net-importer-of-rare-earths-in-2018-101393333.html. Accessed 25 Mar 2021.

Li, Yifei, and Judith Shapiro. 2020. *China goes green: Coercive environmentalism for a troubled planet*. Cambridge: Polity Press.

Liboreiro, Jorge. 2022. Ukraine war: EU urges China to stay neutral but secures no guarantees. *Euronews*, 1 April. https://www.euronews.com/my-europe/2022/04/01/ukraine-war-eu-urges-china-to-stay-neutral-but-without-securing-any-guarantees. Accessed 1 Aug 2022.

Neo Performance Materials. 2021. Energy fuels and neo performance Materials rare earth shipments. *Neo Performance Materials*, 7 July. https://www.neomaterials.com/energy-fuels-and-neo-performance-materials-contract/. Accessed 1 July 2022.

———. 2022. Neo Performance Materials To Receive First-Ever Grant Under Europe's Just Transition Fund For Neo's Planned Sintered Rare Earth Magnet Manufacturing Plant In Estonia. *Neo Performance Materials*, 9 November. https://www.neomaterials.com/neo-performance-materials-to-receive-first-ever-grant-under-europes-just-transition-fund-for-neos-planned-sintered-rare-earth-magnet-manufacturing-plant-in-estonia/. Accessed 15 Nov 2022.

New China TV. 2021. Cooperation under BRI makes Progress, Show Resilience amid Pandemic: Chinese FM. *New China TV*, 2 January. https://www.youtube.com/watch?v=h83fhuxcPL8. Accessed 1 Sept 2022.

Nikkei Asia. 2021. How it happened: Transcript of the US-China opening remarks in Alaska. *Nikkei Asia*, 22 March. https://asia.nikkei.com/Politics/International-relations/US-China-tensions/How-it-happened-Transcript-of-the-US-China-opening-remarks-in-Alaska. Accessed 10 Sept 2022.

Norris, William J. 2016. *Chinese economic statecraft: Commercial actors, grand strategy, and state control*. Cornell University Press.

NPR. 2021. The senate passes a bill to encourage tech competition, especially with China. *NPR.org*, 8 June. https://www.npr.org/2021/06/08/1004600001/the-senate-passes-a-bill-to-encourage-tech-competition-especially-with-china. Accessed 12 June 2021.

Peters, Heidi M., et al. 2022. 2022 Invocation of the defense production act for large-capacity batteries: In brief. *Congressional Research Service*, 27 May. https://crsreports.congress.gov/product/pdf/R/R47124. Accessed 1 Aug 2022.

Piore, Adam. 2022. Beijing's plan to control the World's data: Out-Google Google. *Newsweek*, 7 September. https://www.newsweek.com/2022/09/16/beijings-plan-control-worlds-data-out-google-google-1740426.html. Accessed 14 Sept 2022.

Pottinger, Matt, and David Feith. 2021. Opinion | China is winning the big data war, Thanks to Xi. *The New York Times*, 30 November. https://www.nytimes.com/2021/11/30/opinion/xi-jinping-china-us-data-war.html. Accessed 11 Sept 2022.

Powell, Jamie. 2021. The European electric vehicle landscape. *Financial Times*, 27 January. https://www.ft.com/content/4cbf843e-2503-462c-a9cf-1fdc502ffddc. Accessed 1 Sept 2022.

Preisinger, Irene and Victoria Bryan. 2018. China's CATL to build its first European EV Battery Factory in Germany. *Reuters*, 9 July. https://www.reuters.com/article/us-bmw-catl-batteries-idUSKBN1JZ11Y. Accessed 1 Sept 2022.

Qian, Zhou and Sofia Brooke. 2022. China merges three rare earths state-owned entities to increase pricing power and efficiency. *China Briefing*, 12 January. https://www.china-briefing.com/news/china-merges-three-rare-earths-state-owned-entities-to-increase-pricing-power-and-efficiency/

REN21. 2022. Renewables 2022 Global Status Report. *REN 21*. https://www.ren21.net/gsr-2022. Accessed 11 Sept 2022.

Reuters Staff. 2019. Australia, U.S., to cooperate on critical minerals, rare earths. *Reuters*, 20 November. https://www.reuters.com/article/us-australia-rare-earths-usa-idUSKBN1XU0NU. Accessed 3 Sept 2022.

Ridgwell, Henry. 2021. EU suspends China trade deal as tensions grow over Xinjiang, Hong Kong. *Voice of America*, 10 May. https://www.voanews.com/east-asia-pacific/voa-news-china/eu-suspends-china-trade-deal-tensions-grow-over-xinjiang-hong-kong. Accessed 26 May 2021.

Rogers, Katies and Cecilia Kang. 2021. Biden revokes and replaces Trump order that banned TikTok. *The New York Times*, 9 June. https://www.nytimes.com/2021/06/09/us/politics/biden-tiktok-ban-trump.html. Accessed 11 June 2022.

Rogers, Katie, and Brad Plumer. 2021. Biden administration moves to fix supply chain bottlenecks. *The New York Times*, 8 June. https://www.nytimes.com/2021/06/08/us/politics/biden-supply-chain.html. Accessed 10 Sept 2022.

Rogers, Jason, David Stringer, and Martin Ritchie. 2019. *China gears up to Weaponize rare earths dominance in trade war*. Bloomberg.Com, 29 May. https://www.bloomberg.com/news/articles/2019-05-29/china-gears-up-to-weaponize-rare-earths-dominance-in-trade-war. Accessed 11 June 2022.

Scheyder, Ernest. (2022) EXCLUSIVE U.S. Bill would block defense contractors from Using Chinese Rare Earths. *Reuters*, 14 January. https://www.reuters.com/business/energy/exclusive-us-bill-would-block-defense-contractors-using-chinese-rare-earths-2022-01-14/. Accessed 2 Sept 2022.

Shambaugh, David, ed. 2020. *China and the world*. Oxford University Press.

SMM. 2022. Average Price of Spodumene Concentrates Broke Through $5,000/mt, and Battery-grade Lithium Carbonate Prices Gained another 1,500 yuan/mt. *SMM*, 8 August. https://news.metal.com/newscontent/101913013/average-price-of-spodumene-concentrates-broke-through-5000mt-and-battery-grade-lithium-carbonate-prices-gained-another-1500-yuanmt/. Accessed 10 Sept 2022.

Sonmez, Felicia. 2021. Biden signs Uyghur forced Labor Prevention Act into Law. *The Washington Post*, 23 December. https://www.washingtonpost.com/politics/biden-uyghur-labor-law/2021/12/23/99e8d048-6412-11ec-a7e8-3a8455b71fad_story.html. Accessed 1 Sept 2022.

Strategic Market Research. 2022. Electric Vehicle Charging Station Market to Reach $142.46 Billion in 2030 with a CAGR of around 31.14% – Rising Sales in EVs, Lack of Electric Vehicle Charging Stations, Electric-Vehicle Smart Charging and Government Funding. *GlobeNewswire* News Room, 26 May. https://www.globenewswire.com/news-release/2022/05/26/2451479/0/en/Electric-Vehicle-Charging-Station-Market-to-reach-142-46-billion-in-2030-with-a-CAGR-of-around-31-14-Rising-Sales-in-EVs-Lack-of-Electric-Vehicle-Charging-Stations-Electric-Vehicle.html. Accessed 5 Sept 2022.

Swanson, Ana, David McCabe, and Jack Nicas. 2020. Trump Administration to Ban TikTok and WeChat From U.S. App Stores. *The New York Times*, 18 September. https://www.nytimes.com/2020/09/18/business/trump-tik-tok-wechat-ban.html. Accessed 5 Sept 2022.

Taipei Times. 2021. Biden touts US as democracy champ, China lashes out. *Taipei Times*, 12 December. https://www.taipeitimes.com/News/front/archives/2021/12/12/2003769444. Accessed 2 Sept 2022.

The Ministry of Foreign Affairs of Peoples Republic of China. 2020. Joint Leaders' statement on the regional comprehensive economic partnership (RCEP), *Ministry of Foreign Affairs of Peoples Republic of China*, 16 November. https://www.fmprc.gov.cn/mfa_eng/wjdt_665385/2649_665393/t1832612.shtml. Accessed 11 June 2021.

The White House. 2021a. Executive Order on America's supply chains. *The White House*, 24 February. https://www.whitehouse.gov/briefing-room/presidential-actions/2021/02/24/executive-order-on-americas-supply-chains/. Accessed 1 June 2022.

———. 2021b. FACT SHEET: Biden-Harris administration announces supply chain disruptions task force to address short-term supply chain discontinuities. *The White House*, 8 June. https://www.whitehouse.gov/briefing-room/statements-releases/2021/06/08/fact-sheet-biden-harris-administration-announces-supply-chain-disruptions-task-force-to-address-short-term-supply-chain-discontinuities/. Accessed 1 June 2022.

———. 2021c. Executive order on addressing the threat from securities investments that finance certain companies of the People's Republic of China. *The White House*, 3 June. https://www.whitehouse.gov/briefing-room/presidential-actions/2021/06/03/executive-order-on-addressing-the-threat-from-securities-investments-that-finance-certain-companies-of-the-peoples-republic-of-china/. Accessed 1 June 2022.

———. 2022a. FACT SHEET: The inflation reduction act supports workers and families. *The White House,* 19 August. https://www.whitehouse.gov/briefing-room/statements-releases/2022/08/19/fact-sheet-the-inflation-reduction-act-supports-workers-and-families/. Accessed 1 June 2022.

———. 2022b. FACT SHEET: Chips and science act will lower costs, create jobs, strengthen supply chains, and counter China. *The White House*, 9 August. https://www.whitehouse.gov/briefing-room/statements-releases/2022/08/09/fact-sheet-chips-and-science-act-will-lower-costs-create-jobs-strengthen-supply-chains-and-counter-china/. Accessed 1 Sept 2022.

U.S. Department of Defense. 2022. Defense Production Act Title III Presidential Determination for Critical Materials in Large Capacity Batteries. *U.S. Department of Defense*, 2 April. https://www.defense.gov/News/Releases/Release/Article/2989973/defense-production-act-title-iii-presidential-determination-for-critical-materi/https%3A%2F%2Fwww.defense.gov%2FNews%2FReleases%2FRelease%2FArticle%2F2989973%2Fdefense-production-act-title-iii-presidential-determination-for-critical-materi%2F. Accessed 17 Sept 2022.

U.S. Department of State. 2019. Energy Resource Governance Initiative (ERGI). *Bureau of Energy Resources,* 6 June. https://www.state.gov/wp-content/uploads/2019/06/Energy-Resource-Governance-Initiative-ERGI-Fact-Sheet.pdf. Accessed 12 Sept 2022.

U.S. Geological Survey. 2021. Mineral commodity summaries 2021. *U.S. Geological Survey*, 29 January. https://doi.org/10.3133/mcs2021. Accessed 10 Sept 2022.

U.S. Senate. 2021. Chairman Menendez Announces Bipartisan Comprehensive China Legislation. *United States Senate Committee on Foreign Relations*, 8 April. https://www.foreign.senate.gov/press/dem/release/chairman-menendez-announces-bipartisan-comprehensive-china-legislation. Accessed 18 Sept 2022.

Vekasi, Kristin. 2019a. Politics, markets, and rare commodities: Responses to Chinese rare earth policy. *Japanese Journal of Political Science* 20 (1): 2–20. https://doi.org/10.1017/S1468109918000385.

———. 2019b. China's control of rare earth metals. *The National Bureau of Asian Research (NBR)* (blog), 13 August. https://www.nbr.org/publication/chinas-control-of-rare-earth-metals/. Accessed 11 June 2021.

Whalen, Jeanne, and Ellen Nakashima. 2021. Biden expands Trump order by Banning U.S. Investment in Chinese companies linked to the military or surveillance technology. *Washington Post*, 3 June. https://www.washingtonpost.com/technology/2021/06/03/investment-ban-chinese-surveillance-tech/. Accessed 26 May 2022.

WTO. 2015. *Dispute settlement – The disputes – DS431*. WTO, 15 May. https://www.wto.org/english/tratop_e/dispu_e/cases_e/ds431_e.htm. Accessed 30 Apr 2022.

Xi, Jinping. 2022. Full text of Xi Jinping's speech at China's (20th) Party Congress. *Bloomberg*, 18 October. https://www.bloomberg.com/news/articles/2022-10-18/full-text-of-xi-jinping-s-speech-at-china-20th-party-congress-2022. Accessed 24 Oct 2022.

Xu, Yifan. 2022. US' One-China Policy eroded by Pelosi Visit. *Global China Daily*, 8 July. https://global.chinadaily.com.cn/a/202208/08/WS62f04ad8a310fd2b29e70c66.html. Accessed 8 Sept 2022.

Sophia Kalantzakos is Global Distinguished Professor in Environmental Studies and Public Policy at New York University Abu Dhabi. Her research centers on the geopolitics of critical minerals, the transition to a net zero future, and the fourth industrial revolution. Her work, in particular, examines how resource competition in an era of fraught geopolitics has tilted the balance toward more securitized assessments of global interdependence. Moreover, she examines China's global aspirations manifested in the belt and road initiative, Europe's reckoning with a seismic push against both its normative and economic power, and the US's re-evaluation of its leadership role in the global order. Kalantzakos' work analyzes how current epistemic systems will need to give way to new modes of thinking. As nation-states are turning inward in response to demands for de-globalization even while future challenges remain intensely global, her work advances the construction of dynamic, inclusive, and action-oriented responses. Her publications include *China and the Geopolitics of Rare Earths* (Oxford University Press, 2018; rev.2021) and The EU, US, and China Tackling Climate Change: Policies and Alliances for the Anthropocene (Routledge, 2017). She was a Rachel Carson (RCC) Fellow at LMU Munich in 2015 and 2018 and has served as a member of its Academic Advisory Board. She is currently the President of the RCC Society of Fellows. Recently she has been a Fung Global Fellow at the Princeton Institute for International and Regional Studies (PIIRS) and Senior Fellow in the Research Institute for the History of Science and Technology at Caltech and the Huntington.

Chapter 2
Interdependence vs. Geopolitics: Securitization and Partial Recoupling of Sino-American Relations

Xiaoyu Pu

Abstract Economic interdependence is traditionally viewed as essential to peace and cooperation in international relations. However, in an era of rising geopolitical competition, broad economic ties have become a new source of tensions and conflicts. This chapter analyzes how strategic competitions and political tensions shape economic and technological ties between the United States and China. There are enduring problems in the relationship. Beijing worries that the U.S. will disrupt China's domestic order, while Washington is concerned about China's challenge to the U.S.-led international order. Security concerns of both countries have shaped their policies toward each other. The scenario of complete decoupling is too costly for both economies. Despite political tensions, the United States and China must be prepared to cope with each other as a major power for a long time. With the framework of competitive coexistence, China and the United States should build a responsible and resilient recoupling scenario. Unlike complete and limited decoupling, recoupling seeks to preserve most normal economic ties between the two countries while carefully managing dependencies and risks.

Keywords China · United States · Interdependence · Decoupling · Recoupling · Geopolitics · Securitization · Competitive coexistence

2.1 Introduction

Four decades after establishing a formal diplomatic relationship, China and the United States face rising tensions related to trade, technology, human rights, and Taiwan (Rudd 2022). Economic relations between the United States and China, involving trillions of dollars in financial and commercial flows, are now at risk as

---

X. Pu (✉)
Department of Political Science, University of Nevada, Reno, NV, USA
e-mail: xpu@unr.edu

S. Kalantzakos (ed.), *Critical Minerals, the Climate Crisis and the Tech Imperium*, Archimedes 65, https://doi.org/10.1007/978-3-031-25577-9_2

political momentum builds toward a potentially confrontational future. Some fundamental ideas, such as engagement, that once guided the relationship are now being called into question (Campbell and Ratner 2018). Many worry that the two countries will be entrapped in a new Cold War. As American politicians are increasingly taking a tough position on China, American public perception of China has become more negative. Both sides wish to reduce vulnerabilities; however, complete decoupling would undoubtedly result in a new Cold War. Given the deep interdependence between the two countries, American and Chinese policymakers would be better advised to handle the bilateral relationship prudently. While a total decoupling is less likely, some degree of adjustment seems to be inevitable.

The chapter proceeds as follows. The first section reviews the benefits and challenges of economic interdependence. The second analyzes how political tensions shape Sino-American relations. The third section examines the scenario of a new Cold War and complete decoupling. The fourth section analyzes the scenario of competitive coexistence and partial recoupling. The conclusion summarizes the findings and implications.

2.2 Rethinking Interdependence

The United States and China are now rethinking their relationship. From the beginning of their diplomatic relations in the 1970s, American and Chinese leaders have always dealt with complex political issues such as the status of Taiwan and human rights. But geopolitical tensions and economic engagement were largely separated in previous years, and both governments could somewhat shield economic ties from fallout due to political tensions. In recent years, intensified competition is reshaping broader economic and technological relations between the two countries.

U.S.-China relations are increasingly defined by a paradox: deep interdependence on the one hand and rising political and security tensions on the other. The political and security tensions increasingly have spillover effects that negatively impact economic and technological ties. Both countries' officials, experts, and pundits are questioning many of the fundamental ideas that once guided the relationship. Economic interdependence is traditionally viewed as essential to peace and cooperation in international relations (Oneal et al. 2003). However, broader economic ties might become a new source of tension and conflict in an era of rising geopolitical competition. While liberals typically emphasize the positive ramifications of economic interdependence for peace and cooperation, realist thinkers are always cautious about the negative ramifications of interdependence. Kenneth Waltz pointed out the potential vulnerability of interdependence a long time ago when he wrote, "Many seem to believe that a growing closeness of interdependence improves the chances of peace. But close interdependence means closeness of contact and raises the prospect of occasional conflict" (Waltz 1979, 138). Dale Copeland demonstrates that when leaders have positive expectations of the future trade relationships, they want to remain at peace to secure economic benefits. When these expectations turn negative, leaders are likely to fear a loss of access to raw materials and markets, giving them more incentive to initiate crises to protect their interests (Copeland 2015). Some scholars now emphasize that great powers can "weaponize" interdependence for coercive purposes (Farrell and Newman 2019, 42–79). In particular, asymmetric networks allow nation-states with effective jurisdiction over the central economic nodes and appropriate domestic institutions to weaponize these structural advantages for coercive purposes. Scholars identify two mechanisms: first, nation-states can employ the "panopticon effect" to gather strategically valuable information; second, they can employ the "chokepoint effect" to deny network access to adversaries (Farrell and Newman 2019, 42–79).

As the primary beneficiaries of economic globalization, China and the United States have benefited from their broad social and economic ties. In the early years of the opening and reform era, China learned from the United States more explicitly. Deng Xiaoping started the opening and reform era during the same period China built a formal diplomatic relationship with the United States. China's opening brought refreshing economic ideas and advanced technology from the United States to China. In the 1990s, Chinese Premier Zhu Rongji used external pressures from the United States as a political incentive to push forward economic reform before China entered the World Trade Organization (Wang 2000). Especially in the economic domain, China has pragmatically learned various ideas from the United States (Zhao 2010). Broad economic and social ties between China and the United States have benefited the latter as well. The United States has expanded its investment and trade with China. Through economic engagement with China, many American companies have increased their profits. Relatively cheaper imports from China have helped lower the cost of living for many American families. In the educational sector, millions of Chinese students have studied in the United States since the late 1970s. As a result, America has trained generations of Chinese elites, including engineers, business leaders, journalists, lawyers, and even some senior officials. Educational exchanges have been beneficial for both China and America. While China has improved its science and technology partially due to the efforts of the returnees from abroad, the United States has benefited from its openness to Chinese students. For instance, the United States has attracted some of the world's most talented students. Many Chinese students choose to stay in the United States after receiving advanced degrees, contributing to America's technological innovation and entrepreneurship. Education has been an important component of U.S. soft power and international influence. Those Chinese students who return to China have typically become new bridge-builders between Chinese and American societies (Li 2005).

While U.S. engagement has partially transformed China from an isolated actor into a major player in the international order, many Americans now complain that such engagement was largely a failure. Some emphasize that economic interdependence has not changed China into a liberal democracy (Campbell and Ratner 2018). To be sure, the U.S. diplomatic opening to China was initially motivated by American strategic interests, and China's political liberalization was not the primary reason for engagement with China (Johnston 2019). But there is no doubt that changing China has always been a part of U.S. policy on the country. In his *Foreign Affairs* article published in 1967, Richard Nixon wrote:

> Taking the long view, we simply cannot afford to leave China forever outside the family of nations, there to nurture its fantasies, cherish its hates and threaten its neighbors.... The world cannot be safe until China changes. Thus, our aim, to the extent that we can influence events, should be to induce change. (Nixon 1967)

Many years later, in advocating bringing China into the World Trade Organization, President Bill Clinton argued:

> I think you ought to do it for two reasons: First of all, our markets are already open to China; this agreement will open China's markets to us. And second, it will plainly advance the cause of peace in Asia and promote the cause of change in China. No, we don't know where it's going. All we can do is decide what we're going to do. But when all is said and done, we need to know we did everything we possibly could to maximize the chance that China will choose the right future. (Johnston 2019)

As Johnston correctly points out, American leaders such as Bill Clinton were not certain that U.S. engagement would change China into a liberal democracy (Johnston 2019). However, many American policymakers did hope that China would become more liberal one way or the other.

While political liberalization and democratization were not the main criteria analysts used for judging the success of the policy, many American politicians and experts did have high hopes for engagement with China. Regarding economic reform, engagement with China was supposed to bring about gradual liberalization of the Chinese economy. American business interests and expectation of China's political liberalization were key factors that drove American decisions to grant China most-favored-nation trading status in the 1990s. These considerations also supported China's accession to the World Trade Organization in 2001. In recent years, China's economic market-driven reform has stalled. Contrary to American expectations, Beijing has strengthened its state capitalist model even as it has gotten richer. Rather than opening the country up, the Chinese Communist Party (CCP) is instead consolidating state-owned enterprises and pursuing industrial policies. From the U.S. perspective, Beijing has resisted pressure from Washington to level the playing field for American companies. Through various industrial policies, Beijing has increasingly restricted market access and forced non-Chinese firms to share technology, while funneling investment and subsidies to state-sponsored domestic players (Pearson et al. 2021).

Economic growth was supposed to bring political liberalization. China's burgeoning middle class citizens were supposed to demand new rights, and pragmatic officials should have embraced domestic reforms that would be necessary for further progress. Even after the Tiananmen crackdown in 1989, many Americans still expected the Chinese government to allow for political liberalization. They believed that the internet revolution of the 1990s would encourage such trends by further exposing Chinese citizens to the world. In contrast to these expectations, Beijing did not embrace the politically positive effects of economic globalization. Instead, Beijing responded to the forces of openness by building the "Great Fire Wall" and tightening government control. The government has increasingly constricted the free flow of people, ideas, and commerce. While many expect that engagement would lead to societal liberalization, a pluralization of preferences and interests within Chinese society, events of the last decade have dashed even modest hopes for political liberalization. Rather than empowering the Chinese people, the internet has strengthened the hand of the Chinese state, helping the CCP control information flows and monitor citizens' behavior (Roberts 2020).

As China has become more authoritarian internally and more assertive internationally, some American strategists worry that U.S. engagement with China has essentially empowered a formidable rival (Mearsheimer 2021). Under such a context, broader social and economic ties between America and China are shaken by political tensions and ideological differences. Regarding political disputes, Beijing increasingly uses its market access to pressure American companies and organizations. However, freedom of expression and human rights are America's core values. Hollywood, the National Basketball Association, and even some American universities are frequently put in the middle of clashes of contrasting political correctness: when accommodating Beijing's political positions, American companies are sometimes criticized in the United States; however, if they defend American values against Beijing's will, American companies might potentially lose their market in China (Cha and Lim 2019).

2.3 Geopolitics and Securitization of Sino-American Relations

How does the Sino-American relationship become securitized? The fundamental challenge is the so-called "two-order problem:" the United States worries that China might challenge the American-led international order, while China worries about American disruption to the CCP's leadership in the domestic order (Wang 2021). Before the current downward spiral in the bilateral relationship, the United States and China maintained an implicit understanding: America would not openly attempt to destabilize China's domestic political order, and in turn, China would not intentionally weaken the American-led international order. With this mutual understanding, the two countries expanded their commercial, educational, and social links. They also started to cooperate on various regional and global issues such as nonproliferation of nuclear weapons and climate change. In recent years, the implicit understanding seems to have unraveled: some of Trump's policies tried to openly weaken the CCP, and China has become more determined to challenge some aspects of the U.S.-led global order (Wang 2021).

The United States is increasingly concerned that China is challenging America's global leadership. The Trump administration's National Security Strategy labelled both China and Russia as revisionist powers (The White House 2017). Largely embracing the competitive approach, the Biden administration's National Security

Strategy identifies China as "the only competitor with both the intent to reshape the international order and, increasingly, the economic, diplomatic, military, and technological power to advance that objective." (The White House 2022). The Biden administration's rhetoric on China has become less belligerent but still reflects an antagonistic mood.

According to power transition theory, the most dangerous scenario is that the rising power is rapidly narrowing the power gap with the established power, and this rising power is unsatisfied with the status quo of the existing order (DiCicco and Levy 1999). There are indicators that the power gap between the United States and China is narrowing. Americans increasingly worry that Chinese officials seem to feel increasingly emboldened. Chinese President Xi Jinping told a group of high-ranking CCP officials, "The world is undergoing profound changes unseen in a century, but time and situation are in our favor" (Xinhua 2021). Chinese top diplomatic official Yang Jiechi publicly asserted that "the United States does not have the qualification... to speak to China from a position of strength" (BBC 2021). However, as China's economic growth slows down, there is a new uncertainty about whether the Chinese power has peaked (Beckley and Brands 2021; Mastro and Scissors 2022). Regardless of China's power trajectory, China and the United States will have a more intensified competition in the coming years.

As politicians perceive China more negatively, the American public also increasingly views China more negatively. According to the Pew Research Center, 73% of Americans polled reported holding a negative view of China (Silver et al. 2020). The general Chinese perception of the United States has also become more negative. According to a public opinion survey in 2019, only 21% of the Chinese public had a favorable opinion of the United States (Wang 2020). A recent study of Chinese opinions indicates that young Chinese have more negative views of the United States than older Chinese respondents(Fang et al. 2022).

China has several concerns and complaints against the United States. In the eyes of CCP elites, American interference in China's internal order will potentially undermine the CCP leadership, and this is viewed as the biggest threat to China's security. According to Peking University professor Wang Jisi, Chinese leaders are especially sensitive to domestic disorder caused by foreign threats (Wang 2011). According to former Chinese diplomat Dai Bingguo, regime security is China's top priority in defining national interest. Following regime security are territorial integrity and the development of the Chinese economy (Dai 2010). Regarding maritime disputes in the South China Sea and the East China Sea, American leaders assume that it is vital to defend America's treaty allies against potential threat from Beijing. However, the Chinese elites see U.S. military presence in Asia largely as a threat to China's regional security interests. While nationalism is an essential source of the CCP's domestic legitimacy, Chinese elites have constantly feared that America could damage China's political unity and stability. Many American politicians tend to think it is crucial for the United States to defend a democratic Taiwan in Asia. However, the Chinese elites typically view American support to Taiwan as a major obstacle for China's national reunification. While American elites generally assume that the United States should defend human rights in places such as Xinjiang and

Tibet, their Chinese counterparts often see the American policy in these areas as the latest example of American disruption of China's internal stability.

In recent years, Beijing has increasingly highlighted and even exaggerated the dysfunctions of the American system for China's domestic consumption. America's social challenges, such as racism and gun violence, have been widely reported by Chinese media. By highlighting the problems in a democratic America, the CCP seems to want to justify the continuity of its current one-party system to the Chinese public. No longer a passive recipient of American criticism on human rights issues, Beijing has already used the human rights concept to criticize America.

China's negative perception of America has also been fueled by the legacy of the Trump administration (Zuo 2021). Trump's trade war against China has strengthened Chinese nationalistic sentiment. Many Chinese now believe that America will not accept China's emergence as a global power. The Chinese elites and public have a diminishing belief that China should be so reliant on America. As Chinese officials worry about technological dependency on the United States, they have started to adjust their policy, emphasizing self-reliance and indigenous innovation. The Chinese propaganda system has occasionally promoted Beijing's tough international image in front of a domestic audience.

Above all, the security concerns of both countries have shaped their policies toward each other. On the Chinese side, the vigilance against perceived U.S. interference shapes China's thinking about its policy toward the United States. To safeguard the CCP leadership, the Chinese government has implemented some new laws and policies aimed at restricting the ability of Americans to encourage political activism against the CCP. Beijing has also stepped up its "political education" among cadres and the public at home. Washington has taken a tougher stand on commercial and technological exchanges with China, and there are some voices calling for commercial and technological decoupling.

2.4 New Cold War and Complete Decoupling

Many people tend to use the analogy of a new Cold War to describe the trajectory of U.S.-China relations(Brands and Gaddis 2021). Like the Soviet Union, China is a strategic competitor with an illiberal political system and global ambitions. China will pose a lasting challenge to the United States. In this sense, the Cold War analogy has intuitive appeal.

In 2020, the Trump administration almost implemented a Cold War policy on China. Trump officials often portrayed China as an existential threat to America and argued that the United States must lead a global democratic alliance to weaken China abroad and to foster fundamental political change within China. The Trump administration intensified military competition with China in the Indo-Pacific region. In 2018, the Trump administration started a trade war against China. The voices for broad-scale economic and technological decoupling between the two countries were loud during the Trump era. The Trump administration put some

Chinese companies on the Commerce Department's so-called export control entities list. The Biden administration has largely continued much of Trump's tough approach, although it implements its China policy in a slightly more nuanced manner. The Biden administration has expanded the Commerce Department's export control entities list to include more Chinese companies (Reuters 2022). In August 2022, President Joe Biden signed into law the CHIPS and Science Act of 2022. This law aims to stimulate research and manufacturing of chips in the US while barring some American companies from building advanced tech factories in China (Lo 2022). In October 2022, the US Commerce Department announced some new export controls to restrict China's ability to obtain advanced computing chips, develop and maintain supercomputers, and manufacture advanced semiconductors (US Department of Commerce 2022).

Despite the aforementioned worrying trends, a new Cold War and complete decoupling are unlikely. Compared with the Soviet Union, China today is much stronger economically and more sophisticated diplomatically. Unlike the Soviet Union, China has maintained broad economic and social ties with the United States, and it is also deeply integrated into the world economy as the largest trading partner for many countries. The Cold War between the United States and the Soviet Union had three essential elements: global ideological struggle, opposing military alliance systems, and divided economic blocs. These elements are still largely missing in Sino-American relations, though there are worrying concerns in specific domains (Christensen 2021).

China has tried to avoid framing relations with the West as a new Cold War. In recent years, as Chinese foreign policy becomes more active and assertive globally, some Chinese strategists have started to worry about international pushback and backlash (Pu and Wang 2018). Chinese leaders believe that Soviet-style ideological expansionism could trigger a backlash that might hinder their country's continued growth. In the 19th Party Congress in 2017, Xi Jinping boldly declared:

> The path, the theory, the system, and the culture of socialism with Chinese characteristics have kept developing, blazing a new trail for other developing countries to achieve modernization. It offers a new option for other countries and nations who want to speed up their development while preserving their independence. (Xinhua 2017a)

Some American analysts view Xi's speech as an indicator of Beijing's intention to export a "China model" abroad (Tobin 2020). However, it should be noted that Xi's subsequent statements after the Party Congress provided a more nuanced view (China Daily 2017). In December 2017, Xi denied that China was exporting a China model, stating, "We do not 'import' foreign models, nor do we 'export' the Chinese model" (Xinhua 2017b). According to Xi Jinping's report to the 20th Party Congress in 2022, Chinese modernization "contains elements that are common to the modernization processes of all countries, but it is more characterized by features that are unique to the Chinese context" (Xinhua 2022b). The Chinese officials in the reform era have always emphasized "Chinese characteristics" when describing their version of modernization and socialism. It is difficult to export such a Chinese model if this model is deeply rooted in Chinese politics, history, and culture.

By emphasizing China is a developing country with Chinese characteristics, Chinese leaders imply that their political model cannot be exported to other countries.

There are not yet opposing military alliance systems between the United States and China, as there were during the Cold War between the Soviet Union and United States. While the Biden administration is mobilizing Western allies and Indo-Pacific partners to balance against China, many of America's allies do not necessarily share the same sense of urgency. These countries have some concerns about China's challenges to liberal values and regional security. But most countries do not want to choose sides between the United States and China. For instance, as an American military ally, South Korea does not always support America on a variety of disputed issues between the United States and China (Cha 2020). In addition, some member states of the Association of Southeast Asian Nations (ASEAN), such as the Philippines and Vietnam, have had severe disputes with China over its actions in the South China Sea, but these countries still do not want to miss opportunities for economic cooperation with China. Many ASEAN countries are unwilling to join a U.S.-led, anti-China coalition even though they are interested in strengthening security cooperation with America. Despite security tensions in the South China Sea, the ten ASEAN members have maintained strong economic ties with China. In 2020, ASEAN became China's largest trading partner. If Washington unilaterally adopts a total decoupling policy toward China, the United States will alienate allies and partners that are too economically dependent on China. Although these allies share some of Washington's concerns about Beijing's policies, they do not view Beijing as an existential threat to their own security. China has become the biggest economic partner for so many countries. A new Cold War would break the transnational production chain and be too costly for American allies in Europe and Asia.

The emergence of two divided economic blocks will be too costly even for the United States (Zhang and Shi 2020). China is no longer simply an assembly site for high-tech goods, it has become a manufacturer of key components in its own right. China has also become specialized in more sophisticated manufacturing, making its value to importers very hard to replicate and relocate (Nye Jr 2020). Moreover, many international companies in China usually maintain strong relationships with dealers and suppliers that have been built over decades. It is extremely difficult for these companies to attain similar level of supply chain integration outside of China. Under such a context, complete decoupling would compel American industries and companies to leave and lose the Chinese market. The huge size of the Chinese market means that large American companies would be unenthusiastic about a trade war and technological decoupling.

Above all, while there are rising concerns about U.S.-China strategic competition, a new Cold War could still be avoided. Some regional and global problems such as economic crises, nuclear proliferation, and the global pandemic would be impossible to solve if the United States and China failed to cooperate. China is deeply integrated with the world economy. A total decoupling from China would disrupt the world economy immensely and be an extremely costly option for both the United States and its allies. Across the Indo-Pacific region, few countries are really interested in joining a U.S.-led, anti-China coalition even though many countries do have various degrees of concern about China's challenges to their own security.

2.5 Competitive Coexistence and Partial Recoupling

The biggest challenge for the United States and China is to avoid the worst-case scenario: a severe new Cold War with complete decoupling. To prevent this, America and China should try to build a new endurable framework of competitive coexistence (Campbell and Sullivan 2019).

Both the United States and China are enduring nations. Despite the many divides between the United States and China, each side will need to be prepared to cope with the other as a major power for a long time. The United States must be humble about its capacity to determine China's long-term directions and developments. Rather than relying on assumptions about China's trajectory, America's long-term China strategy should be durable whatever the future brings for China. The United States should seek to achieve not a definitive end state but a steady state of coexistence on terms favorable to American interests and values (Campbell and Sullivan 2019). The Chinese side should not underestimate the endurance of American power and resilience. While the United States might not be as dominant as it was before, it will remain the most powerful nation in many respects (Wang 2021).

America and China must manage rising competition carefully. For years, the United States and China seemed to focus on risk mitigation (Medeiros 2021). They both adopted narratives and policies that sought to downplay disagreement and competition while seeking to expand cooperation. In contrast to the previous administrations, the Trump administration not only tolerated friction but actively promoted it. China under Xi Jinping is also more willing to accept friction and risk in defending and advancing China's interests. Experts from both countries are now calling for new thinking on risk management (Medeiros 2021; Zuo 2021). While expecting and tolerating friction and tension, Washington and Beijing should manage certain risks and costs more carefully. The United States and China should build a policy framework that reflects the competitive nature of the relationship but also builds durable ties that can withstand the tensions (Medeiros 2021).

To build the framework of competitive coexistence, partial recoupling in some domains is needed. Partial recoupling refers to the process of adjusting economic ties to build a more sustainable framework. The key feature of partial recoupling is to build a framework of economic ties that could be politically sustainable in the long run. This approach seeks to carefully manage flows of goods and capital between the two countries while preserving normal exchanges in most domains. Some strategists see decoupling as America's unilateral approach to punishing China (Scissors 2020). However, it is obvious that partial recoupling is different from the notion of decoupling. Even a limited version of decoupling still assumes the bilateral relationship is always moving in the direction of reducing interactions.

Given the broad social and economic ties, limited decoupling does not always reflect real trends in the bilateral economic relationship. For instance, despite the Sino-American trade war, the pandemic has made Americans more reliant on imports of Chinese manufacturing products (Hessler 2021). The United States and China have signed a landmark audit deal that allows US regulators to vet Chinese accounting firms while giving China the chance to retain access to the US capital markets (Murdoch et al. 2022). It should be noted that Biden administration officials have refrained from using "decoupling" to describe their overall economic approach toward China. Instead, United States Trade Representative *Katherine Tai* used "recoupling" to describe the Biden administration's trade policy toward China (Lynch 2021). Admittedly in policy practice, partial recoupling might still be similar to limited decoupling or partial disengagement in some circumstances (Friedberg and Boustany Jr 2020).

In a responsible recoupling scenario, both Washington and Beijing should prudently manage economic and technological dependence on each other. The United States is eager to increase its economic competitiveness and reduce vulnerability and dependency. The Biden administration has continued some of Trump's policies, which aim to reduce U.S. dependency on China. This policy stance is driven by both domestic politics and national security concerns. From Trump to Biden, American politicians tend to accommodate domestic political pressure against economic globalization. The United States must also address vulnerability concerns related to the security ramifications of economic interdependence with China. In the past several decades, the U.S. and Chinese economies have evolved in ways that could complicate the ability of the United States to sustain defense production in the event of a military conflict involving the two countries. In areas such as rare earth metals and printed circuit boards, American manufacturers now depend heavily on imports from China. American officials and strategists are rethinking some of these dependencies (Friedberg and Boustany Jr 2020). While some instances of import dependence give legitimate grounds for concern, there is also a risk that claims of the essentiality of various sectors to defense can be deployed to rationalize protectionism. Instead of carelessly invoking the national security provisions in U.S. law, policymakers should proceed systematically to identify areas where dependence on Chinese imports could create bottlenecks (Friedberg and Boustany Jr 2020). Internationally, the Biden administration has strengthened its cooperation with its allies and partners. Launching the Indo-Pacific Economic Framework (IPEF), the Biden's new regional economic initiative has identified four pillars of cooperation, one of which emphasizes the resilience of supply chains (Martin 2022).

On the Chinese side, Beijing is rethinking economic interdependence with the United States (Gewirtz 2020). In response to trade restrictions and limited technological decoupling, the Chinese government is openly espousing a dual economic circulation strategy to build resilience, even while it continues to champion deepening globalization and free trade (Lin and Wang 2021; Xinhua 2022a). To counter U.S. tactics, China will promote innovation through additional capital investment so as to strengthen weak links in supply chains, develop core technologies, and prevent being choked out by the United States (Zuo 2021). China will also strengthen its ability to retaliate against the United States. China has carefully studied the relevant laws and regulations of the West and learned to build an American-style sanctions system, which led to the establishment of its own "Unreliable Entities List" in May 2019 (Zuo 2021). On June 8, 2019, China announced the establishment of its national technological security management list system to counter American supply interruptions and technology restrictions against Chinese high-tech enterprises (Zuo 2021). China has also strengthened its supervision of American companies, which has enhanced its ability to impose sanctions in the future. In the future, China and the United States will likely continue to spar in the high-tech field. Beijing has long been thinking about a "Plan B" to avoid being choked out by the United States. In September 2022, the CCP's Central Commission for Comprehensively Deepening Reform adopted the guideline on improving a new system for mobilizing resources nationwide to achieve breakthroughs in core technologies in key fields under the socialist market economy (Xinhua 2022a).

Despite some restrictions being inevitable, Washington and Beijing should still have rational incentives to maintain most normal economic transactions and social ties. It is in America's interests to lower barriers to importing low-tech consumer products such as clothing and furniture in return for China dropping restrictions on U.S. agricultural products (Friedberg and Boustany Jr 2020). At this point, however, trying to exclude every Chinese-manufactured computer and integrated circuit from the American economy by imposing a blanket ban would be counter-productive. The United States and China need to reaffirm that their economic ties have been largely beneficial for both countries. As for academic exchanges, both America and China must reassure that students and scholars should be allowed to conduct study and research in both countries. In the United States, it is important to recognize that the vast majority of Chinese students and scholars have legitimate reasons to study and research in American universities. The American government should handle any problematic cases on a case-by-case scenario rather than through racial profiling and politically motivated actions. Facing criticism from legal professionals, civil rights groups, and the Chinese American community, the U.S. Justice Department has ended its controversial "China initiative" (Lucas 2022). The Chinese government should also preserve broad social and educational ties between America and China. Tightening censorship in Chinese higher education is counter-productive to China's own innovation and soft power.

At the international level, America and China will seek to mobilize international support in various regions, but the rest of the world will not be interested in joining either of the two opposing blocs. Many European countries share similar values and concerns with the United States, but they do not have a clear consensus regarding the urgency and priorities to deal with their complicated relationship with China. In the Indo-Pacific region, many American allies are concerned about China's intentions and capacity to build a China-centered regional hegemony. However, most of these countries, including American allies such as Japan and South Korea, are the largest trading partners of China. While resenting U.S. hegemony on some issues, China is not organizing a coherent anti-United States club. As a developing country, China will focus on areas in which it feels it has an advantage, such as poverty reduction, trade, international infrastructure, and so on. China's "developing country" label is meant to signal Beijing's geopolitical alignment: even if China catches up with the West economically, China will still lie with the developing world politically (Yan 2021).

Above all, leaders from both countries cannot ignore domestic political pressure and national security concerns. Therefore, the key challenge for the US and China is how to manage their competition prudently without being trapped in zero-sum thinking (Weiss 2022). Both China and the United States must manage the adjusting process carefully and build a sustainable new framework. Through such a framework of partial recoupling, China and the United States will reduce vulnerability while preserving as many social and economic ties between the two countries as they can.

2.6 Conclusion

The Sino-American relationship has changed in fundamental ways. Both the United States and China have benefited from economic globalization, but their economic ties are shaken by political tensions and backlash. Through engaging China, the United States was hoping that China would become more liberal internally and more moderate internationally. In recent years, China has become more authoritarian internally, and it has also started to challenge American global leadership in some domains.

There are enduring problems in the relationship. China worries that U.S. intervention will disrupt China's domestic political order, while America is concerned about China's challenge to the U.S.-led international order. The competing visions of a CCP-led domestic order and U.S.-led international order will continue shadowing the relationship. The rising concerns from both sides have spillover effects. Previously, economic ties and political disputes were largely separated. In recent years, both sides have taken actions to safeguard their preferred national security, and these measures and policies negatively impact broad social and economic exchanges between the two countries.

Facing these worrying trends, some people believe the United States and China will be entrapped into a new Cold War with the looming prospect of complete decoupling. However, both sides should still have rational incentives to avoid such a new Cold War scenario. China today is a strong economic superpower with deep economic ties with the outside world, including with the United States. Cold War elements such as global ideological struggle, opposing military alliances, and divided economic blocs are still largely missing in Sino-American relations.

To avoid a new Cold War, Washington and Beijing must accept two fundamental realities. The first is that despite challenges at home, the CCP's rule in China will likely remain for the foreseeable future. American attempts to change the Chinese political system are likely to backfire by inflaming anti-American sentiment in China. Ultimately China's political future will be shaped by the Chinese themselves.

The second reality is that America will remain the most powerful nation in shaping the global order. Despite all the problems, the United States has some enduring strengths, including its military power, a culture of innovation, and a dynamic society. As some essential components at the heart of the U.S.-China relationship have changed, it is not easy to reset the bilateral relationship in the foreseeable future. President Biden correctly identifies domestic renewal as his top priority. From an American perspective, the best way to promote democracy is to make democracy work at home. From a Chinese perspective, China has also faced enormous domestic challenges including an aging population, environmental degradation, and social inequality. While the US-China relationship will remain competitive in nature, it does not have to be confrontational. American and Chinese societies are both resilient societies. The two great nations must be prepared to compete and co-exist with each other for a long time.

With the framework of competitive coexistence, China and the United States should seek to shape a responsible recoupling scenario. It should be emphasized that partial recoupling is different from decoupling. A responsible approach to partial recoupling will seek to preserve most normal exchanges of goods, ideas, information, and people-to-people ties between the two countries. This approach seeks to carefully manage dependency on each other while reducing tensions and risks.

References

BBC. 2021, March 19. US and China trade angry words at high-level Alaska talks. https://www.bbc.com/news/world-us-canada-56452471. Accessed 01 November 2022.

Beckley, M., and H. Brands. 2021, October 1. The end of China's rise: Beijing is running out of time to remake the world. Foreign Affairs. https://www.foreignaffairs.com/articles/china/2021-10-01/end-chinas-rise. Accessed 01 November 2022.

Brands, H., and J.L. Gaddis. 2021. The new cold war. *Foreign Affairs* 100 (6): 10–20.

Campbell, K., and E. Ratner. 2018. The China reckoning: How Beijing defied American expectations. *Foreign Affairs* 97 (2): 60–70.

Campbell, K., and J. Sullivan. 2019. Competition without catastrophe: How America can both challenge and coexist with China. *Foreign Affairs* 98 (5): 96–110.

Cha, V. 2020. Allied decoupling in an era of US–China strategic competition. *The Chinese Journal of International Politics* 13 (4): 509–536.

Cha, V., and A. Lim. 2019. Flagrant foul: China's predatory liberalism and the NBA. *The Washington Quarterly* 42 (4): 23–42.

China Daily. 2017, December 21. Highlights of Xi's speech at World Political Party Dialogue. https://www.chinadaily.com.cn/china/2017-12/01/content_35161658.htm. Accessed 01 November 2022.

Christensen, T. 2021, March 24. There will be no new cold war: The limits of US-Chinese competitions. *Foreign Affairs*. https://www.foreignaffairs.com/articles/united-states/2021-03-24/there-will-not-be-new-cold-war.Accessed22Sep2021. Accessed 01 November 2022.

Copeland, D.C. 2015. *Economic interdependence and war*. Princeton University Press.

Dai, B. 2010. Adhere to the path of peaceful development. *Beijing Review*. http://www.bjreview.com.cn/document/txt/2010-12/24/content_320851.htm. Accessed 01 Nov 2022.

DiCicco, J., and J. Levy. 1999. Power shifts and problem shifts: The evolution of the power transition research program. *Journal of Conflict Resolution* 43 (6): 675–704.

Fang, S., X. Li, and A.Y. Liu. 2022. Chinese public opinion about US–China relations from trump to Biden. *Chinese Journal of International Politics* 15 (1): 27–46.
Farrell, H., and A. Newman. 2019. Weaponized interdependence: How global economic networks shape state coercion. *International Security* 44 (1): 42–79.
Friedberg, A., and C. Boustany Jr. 2020. Partial disengagement: A new US strategy for economic competition with China. *The Washington Quarterly* 43 (1): 23–40.
Gewirtz, J. 2020. The Chinese reassessment of interdependence. *China Leadership Monitor* 64: 1–16.
Hessler, P. 2021. The rise of Made-in-China diplomacy, *New Yorker.* https://www.newyorker.com/magazine/2021/03/15/the-rise-of-made-in-china-diplomacy. Accessed 01 November 2022.
Johnston, A. 2019. The failures of the 'failure of engagement' with China. *The Washington Quarterly* 42 (2): 99–114.
Li, C., ed. 2005. *Bridging minds across the Pacific: U.S.-China educational exchanges, 1978–2003.* Lexington Books.
Lin, Y., and X. Wang. 2021. Dual circulation: A new structural economics view of development. *Journal of Chinese Economic and Business Studies*: 1–20.
Lo, K. 2022, September 7. US Chips Act bars American companies in China from building 'advanced tech' factories for 10 years. *South China Morning Post.* https://www.scmp.com/news/china/article/3191596/us-chips-act-bars-american-companies-china-building-advanced-tech. Accessed 01 November 2022.
Lucas, R. 2022, February 23. The Justice Department is ending its controversial China Initiative. NPR. https://www.npr.org/2022/02/23/1082593735/justice-department-china-initiative. Accessed 01 November 2022.
Lynch, D.J. 2021, October 4. White House calls for 'new course' on China trade ties. *Washington Post.* https://www.washingtonpost.com/us-policy/2021/10/04/tai-biden-china-trade/. Accessed 01 November 2022.
Martin, E. 2022, September 7. US to start economic talks with 13 nations to counter China. Bloomberg.Com. https://www.bloomberg.com/news/articles/2022-09-07/us-starts-broad-economic-talks-with-13-nations-to-counter-china. Accessed 01 November 2022.
Mastro, O.S., and D. Scissors. 2022, August 22. China Hasn't reached the peak of its power: Why Beijing can afford to bide its time. *Foreign Affairs.* https://www.foreignaffairs.com/china/china-hasnt-reached-peak-its-power. Accessed 01 November 2022.
Mearsheimer, J.J. 2021. The inevitable rivalry. *Foreign Affairs* 100 (6): 48–58.
Medeiros, E. 2021, March 17. How to craft a durable China strategy. *Foreign Affairs.* https://www.foreignaffairs.com/articles/united-states/2021-03-17/how-craft-durable-china-strategy. Accessed 01 November 2022.
Murdoch, S., X. Yu, S. Shen, and M. Price. 2022, August 26. U.S. and China reach landmark audit deal in boon for Chinese tech companies. Reuters. https://www.reuters.com/markets/asia/china-asks-firms-auditors-prepare-us-checks-hong-kong-sources-2022-08-26/. Accessed 01 November 2022.
Nixon, R. 1967. Asia after Vietnam. *Foreign Affairs* 46 (1): 111–125.
Nye, J., Jr. 2020. Power and interdependence with China. *The Washington Quarterly* 43 (1): 7–21.
Oneal, J.R., B. Russett, and M.L. Berbaum. 2003. Causes of peace: Democracy, interdependence, and international organizations, 1885–1992. *International Studies Quarterly* 47 (3): 371–393.
Pearson, M., M. Rithmire, and K.S. Tsai. 2021. Party-state capitalism in China. *Current History* 120 (827): 207–213.
Pu, X., and C. Wang. 2018. Rethinking China's rise: Chinese scholars debate strategic overstretch. *International Affairs* 94 (5): 1019–1035.
Reuters. 2022, August 23. U.S. adds seven China-related entities to export control list. Reuters. https://www.reuters.com/markets/us/us-adds-seven-china-related-entities-export-control-list-2022-08-23/. Accessed 01 November 2022.
Roberts, M.E. 2020. Resilience to online censorship. *Annual Review of Political Science* 23 (1): 401–419.

Rudd, K. 2022. *The avoidable war: The dangers of a catastrophic conflict between the US and XI Jinping's China*. Public Affairs.

Scissors, D. 2020. Partial decoupling from China: A brief guide. American Enterprise Institute. https://www.aei.org/research-products/report/partial-decoupling-from-china-a-brief-guide/ Accessed 01 November 2022.

Silver, L., K. Devlin, and C. Huang. 2020. Unfavorable views of China reach historic highs in many countries. Pew Research Center. https://www.pewresearch.org/global/2020/10/06/unfavorable-views-of-china-reach-historic-highs-in-many-countries/. Accessed 01 November 2022.

The White House. 2017. *The National Security Strategy*. https://trumpwhitehouse.archives.gov/wp-content/uploads/2017/12/NSS-Final-12-18-2017-0905.pdf. Accessed 01 November 2022.

———. 2022. *National Security Strategy*. https://www.whitehouse.gov/wp-content/uploads/2022/10/Biden-Harris-Administrations-National-Security-Strategy-10.2022.pdf. Accessed 01 November 2022.

Tobin, D. 2020. How Xi Jinping's new era should have ended US debate Beijing's ambitions. https://www.csis.org/analysis/how-xi-jinpings-new-era-should-have-ended-us-debate-beijings-ambitions. Accessed 01 November 2022.

US Department of Commerce. 2022. *Commerce implements new export controls on advanced computing and semiconductor manufacturing items to the People's Republic of China (PRC)*. https://www.bis.doc.gov/index.php/documents/about-bis/newsroom/press-releases/3158-2022-10-07-bis-press-release-advanced-computing-and-semiconductor-manufacturing-controls-final/file. Accessed 01 November 2022.

Waltz, K. 1979. *Theory of international politics*. New York: McGraw-Hill.

Wang, Y. 2000. China's domestic WTO debate internal discord over the role of foreign investment in the economy contributed to the slow pace of China's WTO negotiations. *China Business Review* 27 (1): 54–63.

Wang, J. 2011. China's search for grand strategy: A rising power finds its way. *Foreign Affairs* 90 (2): 68–79.

———. 2020. Summary of survey report on mutual perceptions between China and the United States. Institute of International and Strategic Studies, Peking University. http://en.iiss.pku.edu.cn/dynamic/scientific/3904.html. Accessed 01 November 2022.

———. 2021. The plot against China? How Beijing sees the new Washington consensus. *Foreign Affairs* 100 (4): 48–57.

Weiss, J.C. 2022. The China trap. *Foreign Affairs* 101 (5): 40–58.

Xinhua. 2017a. Full text of Xi Jinping's report at 19th CPC National Congress. http://www.xinhuanet.com/english/special/2017-11/03/c_136725942.htm. Accessed 01 November 2022.

———. 2017b. China will not "export" Chinese model: Xi, http://www.xinhuanet.com/english/2017-12/01/c_136793833.htm. Accessed 01 November 2022.

———. 2021. Xi Focus: Xi stresses good start for fully building modern socialist China. http://www.xinhuanet.com/english/2021-01/11/c_139659492.htm. Accessed 01 November 2022.

———. 2022a. Xi stresses improving system for achieving breakthroughs in core technologies in key fields-Xinhua. https://english.news.cn/20220907/4f53722ecf8542d0b5e5d0fe5fb02cc5/c.html Accessed 01 November 2022.

———. 2022b. Full text of the report to the 20th National Congress of the Communist Party of China. https://english.news.cn/20221025/8eb6f5239f984f01a2bc45b5b5db0c51/c.html Accessed 01 November 2022.

Yan, X. 2021. Becoming strong: The new Chinese foreign policy. *Foreign Affairs* 100 (4): 40–47.

Zhang, J., and S. Shi. 2020. The US unanticipated costs of decoupling from China. *China & World Economy* 28 (6): 1–27.

Zhao, S. 2010. The China model: Can it replace the Western model of modernization? *Journal of Contemporary China* 19 (65): 419–436.

Zuo, X. 2021. The Trump effect: China's new thoughts on the United States. *The Washington Quarterly* 44 (1): 107–127.

Xiaoyu Pu is an Associate Professor of Political Science at the University of Nevada, Reno. His teaching and research interests include Chinese foreign policy, emerging world powers, and international relations. He is a member of the Public Intellectuals Program of the National Committee on United States-China Relations (NCUSCR). He has also received fellowships from the Inter-American Dialogue in Washington, DC, Fundação Getulio Vargas (FGV) in Brazil, and the China and the World Program at Princeton University. He is the author of *Rebranding China: Contested Status Signaling in the Changing Global Order* (The Studies in Asian Security Series, Stanford University Press, 2019). His research has appeared in *International Security, International Affairs, The China Quarterly*, and *The Chinese Journal of International Politics*. He is an editorial board member of *The Chinese Journal of International Politics* and *Foreign Affairs Review* (Beijing). He received BA and MA degrees from Nankai University, another MA from Kent State University, and a PhD from Ohio State University. He has been invited to lecture at many universities and think tanks, including the Chinese Academy of Social Sciences, Harvard, Oxford, NATO Defense College, National University of Singapore, and Tsinghua University, among others.

Chapter 3
Securing Supply Chain Resiliency for Critical Rare Earth Metals

Kristin Vekasi

Abstract Critical minerals, including rare earths, face a rapidly increasing demand, but the global supply chain lacks key indicators of resilience including geographic diversity and the ability to withstand disruptions. This chapter outlines and compares how state intervention and industrial policy has been key to building rare earth supply in the United States, China, and Japan and analyzes key indicators of risk for countries that wish to use industrial policy to diversify the rare earth supply moving forward. After examining protection of property rights and political freedom in countries that host existing and probable rare earth mining sites, the chapter argues that countries should invest in liberal democracies with strong property rights protection and pay to internalize the risks *and* rewards of a resilient supply chain. If the rich democracies are not willing to bear the cost of domestic mining and processing themselves then trade with (and dependence on) China is not only key but is the only possible path forward.

Keywords Rare earths · Supply chain security · Resilience · Japan · China · United States

3.1 Introduction

The production of rare earths, like that of many other minerals considered critical, are highly geographically concentrated. As the world is dependent on the Democratic Republic of Congo for cobalt, and South Africa for platinum, the world depends on China for most rare earth mining, and more so for the post-mining processes that turn the minerals into alloys usable in high-end manufacturing. Rare earths came into close focus by the national security community and political risk analysts in

---

K. Vekasi (✉)
Department of Political Science and School of Policy and International Affairs, University of Maine, Orono, ME, USA
e-mail: kristin.vekasi@maine.edu

S. Kalantzakos (ed.), *Critical Minerals, the Climate Crisis and the Tech Imperium*, Archimedes 65, https://doi.org/10.1007/978-3-031-25577-9_3

2010–2011 after the Chinese allegedly placed an export ban on Japanese imports in the midst of a territorial dispute. Prices soared in a brief market panic. Although a feared global shortage of rare earths never materialized, following this perceived use of economic coercion, policymakers in the United States, Europe, and Japan debated different legislative or administrative approaches to establish sources outside of China and protect against supply chain risks. Japan in particular swiftly moved to diversify with some success, but legislation and industrial policy in other countries aimed at diversifying supply or revitalizing domestic capacity was considerably slower (Gholz and Hughes 2021; Vekasi 2019). Attempts at legislation to reinvigorate the American rare earth industry and private sector efforts to reopen mines faltered, particularly as global prices stabilized.

At the same time, critical minerals face a rapidly increasing demand arising from increased production of high-tech products and by the need for rare earths and other minerals in the energy transition. These minerals are necessary elements in green energy technologies such as wind turbines, solar panels, geothermal energy, and electric vehicle among other products (Imholte et al. 2018; International Energy Agency 2021). A World Bank report estimates that the demand for molybdenum and neodymium for green technologies will increase from 100–250% by 2050 over the current baseline (Hund et al. 2020: 72). This demand is on top of products central to national security concerns such as guided missiles, satellite surveillance technology, and more (Humphries 2019; Nassar et al. 2020). In short, it is well-documented that the world will need more rare earths in order to mitigate the risks posed by a changing climate, maintain national security technologies, and simply to compete economically in sunrise sectors. As China's rare earth industry has become more sophisticated and difficult to compete with in either mining or post-processing, demand for rare earths is projected to greatly rise and the world is repeating the same debates and maybe also the same mistakes about supply risks as we did over a decade ago.

Even prior to the 2010 crisis, countries that rely on rare earths for technological development have tried various public policy mechanisms to have more resilient access to these critical minerals. Policies aim to secure a supply that can meet current and future demand, is sufficiently geographically diverse to withstand disruptions, and is untroubled by local protest. Efforts to date include industrial policy that focus on developing alternatives to rare earths, recycling technologies, and the diversification of mining as well as investment in research and development, either in the private sector or in education and basic research. The United States, Japan, the European Union, and China have all incorporated elements of these policies into their budgets, and yet by most measures of resilience including national diversification, these efforts have been at best minimally successful.

This chapter outlines those efforts, focusing particularly on the challenges and opportunities in diversifying supply, and then assesses possible paths forward. The chapter begins by describing the political economy of the rare earth market, focusing on the role of industrial and trade policy. China holds a strong competitive advantage along the value chain, from mine to consumer product largely as a result of successful policy. If China were to "weaponize" rare earths by restricting trade in the context of an economic or foreign policy dispute, producers along the high-tech and green technology supply chain outside of China would face severe shortages of key inputs, at least in the short-medium term (the next 5–15 years). Rare earths are a choke point in the global supply chain, and thus a potential point of vulnerability. Tracing the policies behind why the rare earth market is concentrated in China, and examining the policies used in Japan and the United States, helps suggest paths to greater independence, including diversification in production and capacity building and talent retention in post-production processes.

States have different approaches to supply chain risks, depending on their unique political economies, institutional, intellectual, and financial capacity, and resource endowments. Barteková & Kemp discuss different national strategies for supply risk in critical minerals, including rare earths, focusing on "distinct national policy styles, national interests, resource endowment and historical experience in tackling supply risk" in the United States, Europe, Japan, China, and Australia (Barteková and Kemp 2016: 153). This chapter builds on their insights and unpacks the different roles of the state and private sector in the rare earth market in the United States, China, and Japan to identify common policy approaches and common pitfalls and suggest approaches to building supply chain resilience.

Building resilience through diversification is difficult. The chapter continues with an analysis of the political risks and challenges in geographic diversification along the supply chain. Most new mining ventures fail, and non-Chinese post-production facilities are limited, struggling, or take years to achieve true operative capacity. To illuminate the opportunities and challenges facing critical minerals in the energy transition, I look at the geographic distribution of mines, and how they have evolved over time, using some standard measures of political risk, including political stability, regime type, and property rights protection. I assess these findings with regards to both social/environmental responsibility and corporate feasibility in diversifying rare earths and make recommendations on how to concurrently manage and balance three goals related to rare earths: building a resilient supply, minimizing environmental and social harm to stakeholder communities, and navigating increasingly contentious international politics and the weaponization of trade.

One key finding is that international cooperation in this sector is absolutely vital to meet future demand, a task that is made more difficult by the rhetoric and politics surrounding China's role in the market. In reporting on rare earths, it is often quoted that during his famous southern tour in the early 1990s, Deng Xiaoping famously said that "the Middle East has oil, and China has rare earths" (see Li 2012 for this quote). This phrase is typically used to ascribe an intentionality of economic nationalism and the specter of coercive economic diplomacy typically using overtly politicized and securitized language (Chu 2010 for an example; Klinger 2018: Chapter 1 for analysis). This politicized language and industrial policies risk limiting what this chapter argues is necessary, which is cross-national cooperation in the sector and increases harms that may be done when states seek to diversify along any stage of the production process.

3.2 Market Structure and the Determinants of Rare Earth Supply

The rare earth market was not diverse historically. A century ago, production was concentrated in India. With the emergence of new technological and economic needs in the postwar era, the United States began to invest in both the mining and processing of rare earths and gained global market dominance. The postwar United States approach to building up this sector was then emulated and effectively adapted by China in the 1980s and beyond along with additional policies not used by the United States, and global production decisively shifted there. The core elements of the policies in both the United States and China, which effectively developed the sector and promoted private sector efforts, included a hospitable regulatory environment for mining, government support for research, development, and education, and coordination between research efforts and the private sector that facilitated vertical integration. China additionally used export quotas and differential pricing policies as part of their broader industrial policy.

Success in the rare earth industry is rooted in policy, not geography. While much has been made of China's near-monopoly control of contemporary extraction and processing, the truth is that rare earths are neither that rare, nor are deposits that concentrated in China (Henderson et al. 2011). Rare earths are not *geologically* rare, but are made so by what Klinger calls a geography that is "contingent" on political and historical factors (Klinger 2018). With sufficient motivation and capacity to implement growth-oriented policy, rare earths could be extracted and processed around the world. There is arguably a "decisive role" for state intervention in the industry (Goldman 2014: 141).

3.2.1 Postwar American Policies

In the 1950s and 1960s United States, there was substantial public investment in rare earth research and development. The original state investments in the rare earth sector were in response to Indian trade restriction on the export of monazite in order to strengthen their global position in that industry. The United States invested heavily in the Mountain Pass Mine in California, where abundant bastnaesite deposits were discovered in 1949, when the experience of the trade frictions were fresh. Under the US policy, publicly-funded research was funneled directly into the private sector. Federal funding created the Ames Laboratory, which developed important new techniques in separating and refining rare earths from monazite sands. A publication called the Rare Earth Information Center (RIC) published work in response to specific problems and questions in the industry helped funnel the insights generated from public dollars into private sector enterprises. Public investments built a public to private pipeline where federally funded research enabled private sector innovation (Goldman 2014).

By the 1970s, state support began to decline rapidly in the United States. Molycorp, the owner of the Mountain Pass Mine, mined and processed most of the world's rare earths but the postwar appetite for state support of the industry in the United States had declined. In the 1980s, as demand for permanent magnets grew, firms in Europe and Japan also started to invest in rare earth technologies. As is described in the next section, however, as cheaper mineral alternatives emerged from China, neither Molycorp nor its potential non-Chinese rivals could compete, and their governments did not incentivize them to do so. By 2002, RIC had disappeared completely, decidedly closing the postwar chapter of US rare earth dominance.

3.2.2 The Chinese Approach

The key to China's contemporary rare earth situation is industrial policy, marked by direct state intervention in domestic and international markets, and investment in technical expertise. The broad goals of Chinese industrial policy have been to hold an internationally competitive advantage in all stages of the rare earth value chain. In doing so, they have built an extensive and vertically integrated rare earth value chain inside China both to meet domestic and global demand. This section outlines policy efforts in research and training, trade and pricing policies, building intellectual property, and attracting foreign firms to China to achieve this level of national market concentration. Chinese policies have shared some similarities with the postwar US approach, but there are also key differences: the Chinese approach has far more intervention in the market.

In order to maintain a large market share, China capitalized on its advantages in lower production costs, used export and production quotas, invested in research, education, and development, and policies aimed at centralizing national control of the industry. Part of the consolidation process for China was building domestic technical capacity through state investments in universities and research institutes. This policy mirrored the earlier American approach. China has two research institutes, opened in 1952 and 1963, that focus on basic research in rare earths, but the commitment to rare earth excellence truly began in the 1980s. In the Reform and Opening Era, state investment increased considerably. In 1980, the "father of Chinese rare earth industry" Xu Guangxian started the National Laboratory of Rare Earth Materials Chemistry and Applications at Beijing University. The same year, the Chinese Society of Rare Earths was founded, and they began publishing the China Rare Earth Information newsletter in 1985, serving a similar role as a conduit as RIC between basic research and market goals. Two government research grant programs also nurtured basic rare earth research in China: the National High-tech R&D Program in 1986 (also known as the 836 Program) and the National Basic Research Program in 1997 (the 973 Program). The 836 Program was initiated Deng Xiaoping and aimed to "to strive to achieve breakthroughs in key technical fields that concern the national economic lifeline and national security" (Chinese Ministry of Science and Technology 2007b). It was renewed under five-years plans through the 2000s. The 973 Program was oriented towards the development of basic research "build up a solid S&T foundation for the sustainable socio-economic development of our nation" and particularly to foster innovation in crucial sectors (Chinese Ministry of Science and Technology 2007a). These programs bore fruit. China's success in training the next generation of rare earth expertise is evident in that China is home to the only scientific journals fully dedicated to rare earth research, and outpaces other countries in training new students (Hurst 2010).

The success of China's strategic and intellectual investments are directly observable by a key measure of intellectual property: patents filed. Figure 3.1 shows the number of rare earth sector-related patents by country from 1994–2021, collected from the Google Patent database that documents patents around the world.[1] In the early 1990s, most rare earth-related patents came out of Japan and the United States (the major contributors to the "other" category are countries in Europe, South Korea, and Russia). China's share of new patents increased from 7% in 1994 to almost 70% by 2020. During this time, the total number of rare earth patents filed from both the United States and China almost doubled, while filings from Japan went down. By

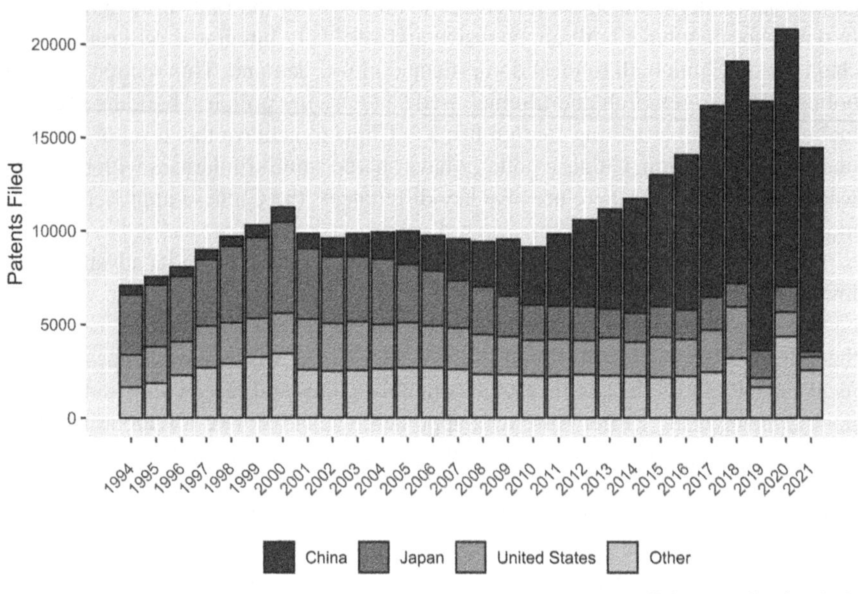

Fig. 3.1 Rare earth patents filed by Country, 1994–2021. (Notes: Data collected and analyzed by author from the Google Patents database)

[1]These data were collected using the Google Patent platform, which collects international patent applications and uses an automatic translation function. The search included patents that referred to "rare earth" or "rare earths". It is likely not exhaustive of rare earth-related patents but is representative of cross-time and cross-national variation.

2008, China had surpassed the United States for patents filed and the same with Japan by 2009. In short, as measured by education, expertise, and the production of intellectual property, China's rare earth policies over the past four decades have been broadly successful, and China's command of global rare earth processes in mining and post-processing have been earned with time, policy, and attention.

China's approach has not been without its downsides. From the 1990s–2000s, the Chinese rare earth industry consisted of many small firms, primarily in Inner Mongolia. As outlined in a 2012 State Council report and independent scholarship, the industry was rife with illegal mining and regulatory violations (Abraham 2015). A study using trade data from Mancheeri and Murakawa comparing the rare earth exports reported by China to the imports from China reported by the rest of the world estimated that between 30–50% of rare earth exports came from illegal mines (N. Mancheri and Marukawa 2016). The environmental and health effects to the communities around these mines were devastating, including poisoned groundwater, high incidences of cancer, and birth defects (Kilby 2014; Klinger 2018).

The policy response to this challenge was centralization of the industry into six regional state-owned enterprises in 2016: Northern Rare Earth, Southern Rare Earth, Chinalco, Xiamen Tungsten, China Minmetals and Guangdong Rare Earth. The consolidation appears to have reduced though not completely eliminated illegal mining or trade (Reuters 2019; Standaert 2019). While the import/export gap in 2010–2014 ranged between 35–55% of total exports, the numbers were much closer to parity in the 2015–2017 period. Trade data from 2018 to 2021, however, shows a return to import-export gaps at around 30% of total exports.[2]

Beyond illegal mining, centralization also served to consolidate the state's control over production quotas and prices, in a pattern long familiar across industries considered strategic by the central government (Hsueh 2011). With a limited number of regionally based mining companies, the state is able to set production quotas, and perhaps even enforce them. The Ministry of Industry and Information Technology sets production, smelting, and separation targets biannually, which affect the overall supply of the minerals in the world. Both mining and smelting/separation targets have increased over the past 5 years, though not steadily (Daly 2021). Production quotas had an additional use to create some artificial scarcity after the limitations of trade policy became clear, as discussed in the next section.

3.2.3 *Trade and Pricing Policy*

Along with abundant reserves and the state investment policies described above, China also used pricing and trade policy to provide advantages to local producers and disadvantages to foreign producers beginning in the 1990s. In the mid-2000s, similar policies were deployed to attract foreign business to China. The Ministry of

[2] These numbers were calculated using rare earth trade data from the UN Comtrade database comparing the difference between exports reported by China and imports reported by partner countries.

Commerce deployed export quotas, differential pricing for domestic and foreign end users of rare earths, and production quotas to create a rare earth infrastructure within China that was far more attractive and efficient for downstream producers than any other market. Export quotas were used from 2005 to 2014, after which China shifted to the production quota system after losing a case from the United States, Japan, and the European Union at the World Trade Organization (World Trade Organization 2015). These direct market interventions aimed to limit supply and increase environmental and safety standards.

Like oil at the height of the OPEC oil cartel, rare earth prices do respond to supply and demand, but China has strong leverage through its commanding market share and price volatility is a source of risk (N. A. Mancheri et al. 2019). In 2009–2011 with feared shortages because of Chinese export quotas as well as geopolitical concerns following the alleged export ban to Japan, prices surged and were extremely volatile. By mid-2013, prices settled at points somewhat higher than they were prior to the late-2010 spikes where they stayed until prices for some alloys began to rise in 2021. Price spikes and volatility have not been consistent across elements or markets. Figure 3.2 shows the price annual price averages for selected elements from 2009 to 2022. Variation across price volatility reveals the nature of single-source dependence risks and more general security risks. While there was volatility across the market in 2010–2011, prices increased far more for foreign buyers than Chinese domestic ones (Argus Media n.d., various years). Sales for Chinese domestic use had significantly less volatility than those for foreign use. Lanthanum oxides, used to produce lenses and lighting equipment, were around

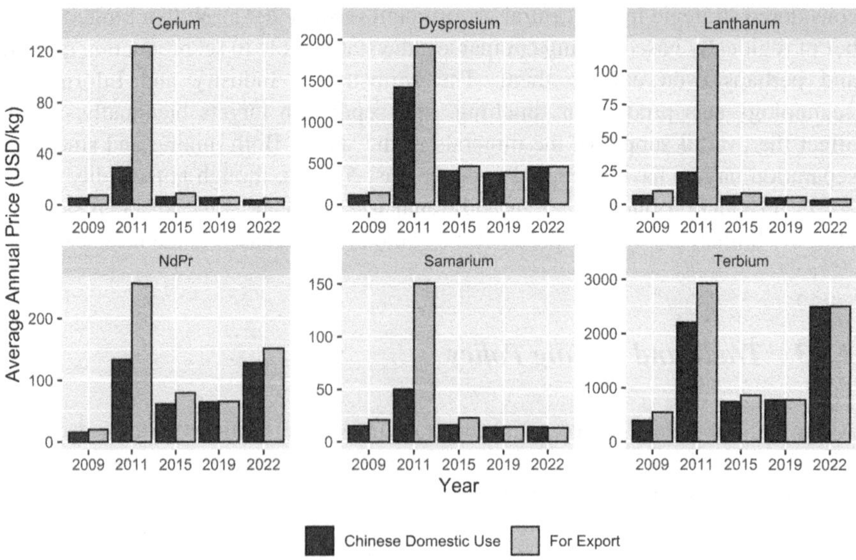

Fig. 3.2 Average annual prices: Selected elements and alloys. (Notes: Price data collected from Argus media; Institute of Rare Earths and Metals)

$2/kg for export and domestic sales prior to 2010. They dramatically spiked to $26/kg domestically (a 23-fold increase) and to $172/kg for export (a stunning 86-fold increase) before falling and remaining at below the 2009 price. Neodymium-praseodymium (NdPr) alloys, a key ingredient in permanent magnets, also rose dramatically in the 2010–2011 period. They never returned to pre-crisis prices and have almost doubled in price from 2019 to 2022. There is also a remaining gap between domestic Chinese prices and those for export. Following from historical experience, if there is price volatility caused by economic sanctions, market intervention, or some other supply crunch, buyers could face extreme volatility and price increases, particularly outside of China.

Reasons behind the price volatility of a decade included the exaggerated response after the alleged export ban from private firms, but were also consequences of Chinese industrial policy. There was a sense of investor panic after feared shortages from China and some companies stockpiled rare earths driving up demand and hence prices. There was real uncertainty about the future of the rare earth market and the degree of export accessibility. Price increases were also driven by the underlying intervention of the Chinese state into the rare earth market: prices were arguably held artificially *low* prior to the panic and then again after it subsided. Even owners of established mines, such as Mountain Pass in California, struggled to survive at these low price points and Molycorp filed for bankruptcy in 2015.

This analysis has two important implications. First, the low prices prior to 2010–2011 were direct results of Chinese market conditions and industrial policy. The flooding of international markets with cheap rare earth metals was one mechanism that allowed China to gain market dominance. Second, after the panic subsided, prices of rare earths largely returned to an equilibrium at the same approximate point where they were prior to the crisis only to start rising again a decade later. As discussed further below, collapsing prices from 2014 were a key mechanism in the failures of diversification efforts by non-Chinese companies: they simply could not achieve solvency.

Chinese pricing policy had a secondary goal beyond outcompeting foreign miners: attracting producers and expertise to China. Two instances of relocation to China demonstrate some success of China's policy. Following the alleged rare earth export ban in 2010, some Japanese companies (for example Hitachi Metals) moved branches of their business reliant on rare earth materials into China in order to take advantage of lower prices and easier access.[3] Following the bankruptcy of the American mining company Molycorp, it was reorganized into Neo Performance Materials (headquartered in Canada) moved post-processing and research and development facilities into China, although they also have a processing facility in Estonia with a much smaller output than the Chinese facilities (Lynch 2019).[4] The

[3] See https://www.hitachi-metals.co.jp/e/corp/corp08.html for a list of Hitachi's overseas locations including six sites in China that work with advanced metals. Hitachi Metals San Huan Magnetic Materials (Nantong) in Jiangsu province opened in 2017.

[4] See also NeoMaterials global sites at https://www.neomaterials.com/wp-content/uploads/2017/11/Snapshot_Unique_Global_Infrastructure-1.jpg

first relocation relates to China's success in using pricing as industrial policy. The second shows the broader success of China's nurture of rare earth talent and infrastructure. While global production is currently rising overall, Chinese partial ability as the swing producer to keep prices low makes it a difficult market for new players to profitably enter. In the 2020s, rare earth prices are once again increasing, driven by industry consolidation and mine closures in China and increasing demand in China and the rest of the world. However, neither new mining nor rising prices can bring expertise in the downstream post-production processes.

3.2.4 The Current Chinese Policy Landscape

Rare earth market watchers and scholars in the 2010s evaluated the efficacy of China's policy on the ability to consolidate the domestic industry, control production numbers and eliminate illegal mining, standardize production procedures, and enforce environmental protections and other regulations. If China was able to successfully implement these policies, then the regime would also be able to control supply to the extent that they could also control prices and increase their industrial edge over the rest of the world in the sector. The 2016 "Rare Earth Industry Development Plan" published by the Ministry of Industry & Information Technology (MIIT) in conjunction with the 13th five-year plan, described many of these policies with specified targets for increased profitability and improvements in the high value-added segments of the industry, meeting higher environmental standards, and decreased production and smelting reflecting the goal of industry consolidation. One goal in the plan was to "improve mechanisms to keep the prices of superior minerals stable through limiting production." The 13th five-year plan, in particular, focused on the shift in China's political economy to higher-value added products with increased environmental sustainability. Goals included strengthening "geological environmental governance and ecological restoration in regions of intensive mineral resource mining" and "green mining" (13th Five-Year Plan 2016). In late 2020 China announced the creation of a new vertically integrated state-owned enterprise called the China Rare Earth Group, which controls the consolidates the southern industry (Zhou & Brooke, 2022). The China Rare Earth Group controls 60–70% of Chinese production (around 30–4% of global supply).

By the time the 14th five-year plan was announced in 2021, many though not all of the previous goals had been met or were in process. While environmental goals have been more elusive in China's growth-oriented political economy, industrial consolidation and price stabilization were quite effective (Economy 2011). China has also decisively moved up the value-added chain, as evidenced by their large research and development investments and expertise in the intermediate stages of production. The 14th five-year plan calls to "promote breakthroughs in advanced metals and inorganic non-metal materials such as high-end rare earth[s]...[to]

accelerate the breakthrough in key technologies". The plan is heavily focused on the newer industrial policy in China to shift towards higher value-added production, green technologies, and an economy more driven by domestic production *and* demand (14th Five-Year Plan 2021). While rare earths are not the only mineral targeted in the plan, these minerals are central to these broader goals. Many of the objectives – electric vehicles, space technology, new materials, and more – will require a reliable source of rare earths for either Chinese producers or foreign producers based in China.

Both the United States and China succeeded in the rare earth industry through state intervention in the market. The US approach focused on investing in basic research that would fulfill the demands of the private sector and continued to expand the market for existing mines, particularly the Mountain Pass Mine. The Chinese approach similarly featured investment in basic research and the development of what has become a state-controlled vertically integrated, world-class industry.

3.2.5 The Japanese Approach

Following the alleged rare earth export ban, Japan assertively pursued international diversification with a combination of industrial policy, diplomacy, and private sector efforts. Japan has no domestic sources and so must look abroad for alternatives. Japan is a state with a history of state guidance and sometimes direct intervention in the development of particular industries and sectors key to economic growth both historically (Johnson 1982), and in the contemporary period (Vogel 2018). As part of the "developmental state" model, Japan's Ministry of International Trade and Industry (now Ministry of Economy, Trade, and Industry or METI) used subsidies, facilitated financing of export-oriented firms, and helped the private sector coordinate in sunrise sectors (Woo-Cumings 1999). Japan also used trade policy to promote exports and limit imports in sunrise industries (Pempel 2021), as well as strategically export sunset industries as foreign direct investment (Solis 2004). Japan thus has policy tools, institutions, and historical experience in implementing these tools of industrial policy.

Japan's historical toolkit of industrial policy informed Japan's approach to dependence on Chinese rare earths, which has been more successful than other countries. Diversification activities included new economic partnership agreements, joint ventures, mining exploration, and rare earth processing plants throughout Asia, the Americas, and Australia. Japan's rare earth diversification efforts have been broad, spanning continents and even investing in efforts to find novel mining locations like the bottom of the Pacific Ocean or space despite the availability of more pragmatically located deposits (Beauford 2011; JOGMEC 2022; Toro et al. 2020). The Japanese state promoted diversification through diplomatic agreements with countries with domestic rare earth reserves, overseas development aid projects, direct subsidies and partnerships with a state-owned enterprise, and by providing opportunities for firms to find overseas partners through economic tours or trade fairs.[5]

The Japanese Ministry of Foreign Affairs pursued partnerships and new agreements in countries with rare earth deposits, but lacked either the capacity, sufficient infrastructure, or domestic demand to safely mine and process. Japan pursued economic diplomacy with the United States, Australia, Mongolia, India, Vietnam and Kazakhstan, including efforts to secure strategic resources through overseas development aid and diplomacy (Maehara 2011). Not all of these efforts were successful. For example, efforts to start new mining in Mongolia and Kazakhstan have largely failed and cooperation with India and Vietnam has been progressing at a slow pace and small scale.

Japan also used industrial policy. The Ministry of Economy, Trade, and Industry introduced competitive grant programs to promote research on recycling technologies and substitution to, in their words, "escape" Chinese rare earth dependence. They also provided grants to companies for their efforts in opening up new international rare earth markets (Ministry of Economy, Trade, and Industry 2010a, b). The recipients of the METI money from industrial policy are largely small or medium-sized enterprises, but the real movers in the sector of critical minerals are large-scale enterprises and the trading companies that help secure their materials, and a state-owned enterprise that provides financial backing to these large companies. As a relatively resource-poor country, Japan established two state organizations in the 1960s to ensure a supply of oil and minerals. In 2004, these organizations were combined into the Japan Oil, Gas, and Metals Corporation (JOGMEC), which is under the jurisdiction of the ministry responsible for most of Japan's industrial policy: METI. While mineral extraction is a key goal, JOGMEC assists along the supply chain. Their goals are to promote early-stage exploration and advanced stage metal mining, helping develop recycling technologies, and and alternatives to rare minerals and metals. They also have a stockpiling program for rare earths. After private companies put in requests for assistance, JOGMEC helps initiate rare earth projects around the world, including in Canada, the United States, South Africa, Australia, Kazakhstan, Vietnam, and Brazil. These are done in partnership with Japanese general trading companies, which are larger companies that (among other roles) solve supply chain problem issues within the Japanese economy for other private firms. For example, JOGMEC and Sojitz are financing the Lynas Rare Earth Project in Australia, which is a key source of non-Chinese rare earths for Japan. With Toyota Tsusho they are helping guarantee a lithium project in Argentina (JOGMEC 2020). While JOGMEC provides financing assistance for these projects, they are initiated by the general trading companies.

Japanese-Australian rare earth-related investments in Malaysia are a successful example of diversification along the supply chain. Lynas runs a rare earth

[5] These policies are outlined in more detail in (Vekasi 2019) and (Gholz and Hughes 2021).

processing plant there with the sediment from their mine in Australia. Lynas was saved from bankruptcy by JOGMEC and Sojitz, enabling their mine and post-processing plant to bring product to market and ensuring a non-Chinese supply of rare earths for Japanese producers, particularly of neodymium and praseodymium used in electric car batteries.

The public-private nexus and use of industrial policy has been key for Japan's efforts in securing a diverse and resilient supply chain. By late 2017, Japan was importing approximately 30% of its rare earths from Asian countries other than China, a trend that has continued through 2021. Many of these come from Malaysia, showing the success of JOGMEC's policies.

Vulnerabilities, however, still remain. Following the onset of the Covid-19 pandemic and severe supply chain disruption in China, Japan initiated new but similarly designed industrial policy to diversify companies from China, either through reshoring or moving into a different overseas market. Through this process, at least three companies have received grants to develop rare earth-related companies in Vietnam and Malaysia (JETRO 2020; Solis 2021). Japan is also pursuing cooperation with its allies in the United States, Canada, the European Union, and Australia to maintain and develop expertise in the rare earth sector. They hold regular meetings to share research and strategies on critical minerals (Ministry of Economy, Trade, and Industry 2021). Japan and the United States also pledged to jointly develop resiliency in critical minerals during a high-level summit in April, 2021 (*Fact Sheet* 2021). Japan has maintained its commitment to state assistance for the private sector in rare earths in the name of supply chain security. Its partial success, combined with observations of the United States and Chinese approach, provide a framework for understanding supply chain resilience in the rare earth sector.

3.2.6 A Framework for Resilience

The complicated and environmentally costly rare earth supply chain rewards countries that invest in comprehensive technological approaches and cushion market actors from costly failures. The United States succeeded in the postwar era due to abundant deposits combined with large government investments in basic research that enabled private firms to pursue technologies more efficiently *within* the firm as they were shielded from costly dead-ends. The Chinese dominance emerged in the 1990s because of abundant deposits, ample state investment in research and development, preferential trade policy, and eventual state consolidation of the industry. Japan managed to somewhat diversify its supply chain due to industrial policy that invested in research and development and diversification and active assistance from a state-owned enterprise. The private sector also matters in building demand, guiding R&D goals, and of course bringing new products to market.

Proactive diversification along the supply chain seems like a clear answer to the double challenges of single-country reliance and rising demand, but includes high environmental, political, and market risks both for companies and the communities that surround existing and potential mines or post-processing plants. In the short term, Chinese price controls make diversification difficult to initiate and sustain without either state intervention or a long-term commitment from heavily capitalized firms. New mining would likely need at least medium-term state subsidization such as we see in the Chinese or Japanese models. With a proper incentive structure that encouraged building towards independent profitability, it could be helpful in the medium and long run, but bears the risk of market distortion and capture from the rare earth industry. The next section outlines possibilities for diversification with regard to political and social risks.

3.3 Risks and Rewards of Diversification

Geographic diversification of both mining and post-production is one possible response to mitigate the risks of dependence on one national source of materials and increase the available supply of rare earths. As 70% of potential rare earth mines lie outside China, geographic barriers are not the primary problem. However, the abundance of non-Chinese deposits obscures the challenges of opening sustainable new mining sites, and moreover of developing non-Chinese capacity for vertical integration. As is outlined in the above discussion, there are many challenges for new production sites, including environmental issues, difficult market competition because of Chinese leverage over prices, length of time to open a mine, lack of expertise to develop a fully vertically integrated supply chain, and political risks such as expropriation or mandated beneficiation. This section evaluates the attempts to diversify mining and post-processing over the past two decade, assessing actual and potential diversification through the lens of the potential environmental, social, and political risks.

Negative environmental externalities in the rare earth industry are not incidental, in either the mining or processing stage. Pollution includes radioactive runoff, mountaintop removal mining, landslides, and groundwater pollution (Ali 2014; Haque et al. 2014). China used the very real environmental damages caused by rare earth mining as the motivation behind trade restrictions (Kilby 2014; Standaert 2019), reasoning ultimately rejected by the World Trade Organization (World Trade Organization 2015). While the WTO decision was ultimately based upon the differential treatment of foreign versus domestic firms, prohibited by China's WTO accession agreement, the environmental concerns were quite real.

Risks also arise from price distortions, particularly the typically low prices of most rare earths. The former American company Molycorp's experience with the California Mountain Pass Mine is instructive. The United States tried to diversify using its domestic reserves. The Mountain Pass Mine had closed in 2002 due to environmental concerns as well as unprofitability (minedat.org 2019). When prices began to rise, and incidentally at the urging of US policymakers, Molycorp reopened the mine in 2012 only to declare bankruptcy in 2015 when prices collapsed to early 2010 levels (Grasso 2013: 36–8).

This sort of business risk is particularly salient for new entrants or smaller, poorly financed firms. In one study of new entrants to the market, analyst James Kennedy found that of 400 publicly listed rare earth start-ups around the world from 2012–2019 only five of them had successfully achieved commercial production. Moreover, he wrote that "of those, only two reached significant volumes. Of those two, one is bankrupt and resurrected with Chinese financing and the other lost its operating permit for a short period" (Kennedy 2019).

A closer look at attempts to open mines throughout Africa is also instructive. There are at least seventeen potential rare earth mining sites in Africa, located in Burundi, Kenya, Tanzania, Malawi, Uganda, Madagascar, Mozambique, South Africa, and Namibia (Harmer and Nex 2016).[6] One was consistently producing rare earths – the Gakara mine in Burundi, which is rich in NdPr, but the Burundi state suspended production in 2021 due to poor financials (Rainbow Rare Earths 2022). Rainbow Rare Earths also runs the Phalaborwa Project in South Africa, which is in final testing stages. These two mines are the vestiges of colonial-era projects that were first exploited in the 1930s in the case of the Malawi Gakara Mine or 1960s in the South African mine. Three of the African mines have failed outright, including the nationalized Mrima Hill Mine in Kenya (ITALaw 2020; Obulutsa and Malalo 2018), a German/Singapore joint partnership in Madagascar (Carver 2017), and the Xiluvo Mine in Mozambique (African Mining 2019). The Makuutu Mine in Uganda – a potential source of heavy rare earths – is an Ugandan/South African/Australian joint venture and is at the final testing stages (Ionic Rare Earths 2022). Others are at very early stages, despite initial investments more than a decade prior. There are issues of contested ownership, nationalization, and of course environmental concerns, which also plague the Malaysian processing plant run by Lynas (Sukumaran 2019).

In short, opening new mines is not a short or simple process. In incentivizing new domestic mining, there needs to be a commitment to carry through short-term price shocks and a commitment to solving problems with local governments and communities in the event of environmental or other issues. The next section analyzes the potential mining sites with respect to these risks.

3.3.1 Potential Diversification

In 2002, the US Geological Survey published a list of 799 current mines and potential deposits for rare earths. Coding these sites by 2002 production status and site history produces the data shown in Fig. 3.3, which sorts by geographic location by continent and status. Current producers are mines that were producing rare earths as of 2002, and past producers, mines that had previously mined rare earths. "Active mines, not rare earths" are mines that in 2002 were actively mining other minerals,

---

[6] The 17 projects were identified from Harmer and Nex (2016), as well as scouring websites and financial statements of mining companies and mining data conglomeration websites.

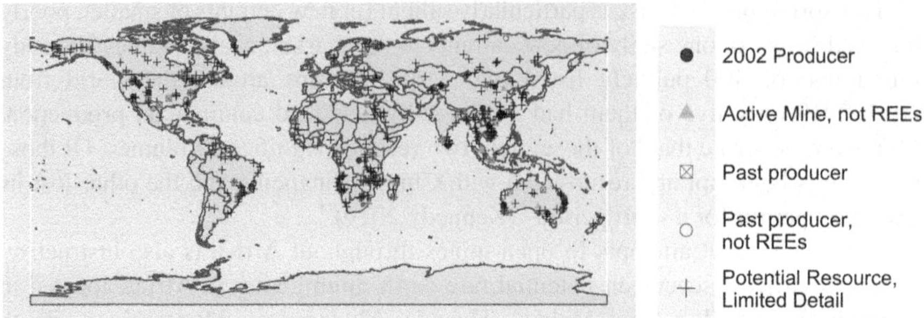

Fig. 3.3 Map of active mines and potential deposits by 2002 Status. (Notes: Data from Orris, Greta J., and Grauch, Richard I., 2002, Rare earth element mines, deposits, and occurrences: U.S. Geological Survey)

and the "past mines, not rare earths" are similarly former mines of other minerals. Altogether, 209 mines were either current or former producers of rare earths or other minerals that could potentially produce usable rare earths as a byproduct. The last categories included sites designated, with various language from the USGS, as potential resources, sites with rare earth occurrences, or contained no details. These sites are by far the most common (74%), showing the lack of detailed global surveying knowledge about rare earth deposits. These data are not weighted by production, so a small producer (or a potential site with small predicted reserves) is counted here as equivalent to a major producer like the Bayan Obo mine of Inner Mongolia. In 2002 when the data were published, only 5% of the potential sites were producing, and some of those mines closed shortly after the data set was published (such as the Mountain Pass mine in California). In 2017, the USGS published a new database of current or highly probable sources of critical minerals, including rare earths (Labay et al. 2017; Schulz et al. 2017). These data, along with more recent information from company websites to confirm the status of mines, identify 58 current sites from which global producers rely, and are the locations upon which global supply will depend in the coming decades as newer projects have high failure rates and take considerable time to move into substantial production. Figure 3.4 shows a map of these 58 sites.

There is broad potential geographic diversity in both deposits and mines, although by scale most production still occurs in China. In 2002, 34% of the potential mines were in Asia, and 28% in the Americas. A majority (54%) of producers, whether of rare earths or other minerals, were located in Asia and most of these are in China. The second most common source was Oceania, largely Australia. The current data shows more geographic diversification in terms of mining sites: 41% are in Asia including a large uptick in Vietnam, 26% in the Americas, 17% in Africa, 10% in Oceania, and 5% in Europe. While this trend is encouraging with respect to both increasing global supply of rare earths and increased resilience from diversification, it does not address other aspects of resilience, including political, environmental, or social risk.

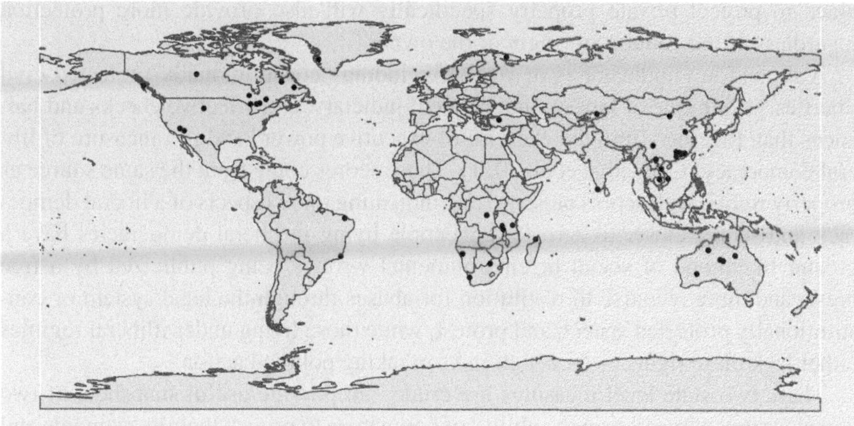

Fig. 3.4 Map of Current and Probable Rare Earth Mines 2021. (Notes: Data from Orris, Greta J., and Grauch, Richard I., 2002, Rare earth element mines, deposits, and occurrences: U.S. Geological Survey; Author's research)

3.3.2 Diversification and Risk

To identify the potential for new rare earth supply chains, it is necessary to also identify locations that have the conditions for resilience: stability and predictability. This section looks at the locations of the 58 existing and probable mines with respect to social and political factors, including property rights and individual freedoms in order to evaluate potential for diversification in the short-to-medium term.

The business decision to move into new markets or explore new mining or processing opportunities is determined by the risk/reward structure of the global market. While foreign investment decisions are largely determined by macroeconomic factors such as exchange rates, market access and proximity, and the costs of labor, political determinants also matter (Dunning 1980). Two factors are particularly important looking at global foreign direct investment decisions: regime type and the strength of property rights (Ahlquist 2006; Jensen 2003). With natural resources, investments also need to follow the reserves, which necessarily limits choice. However, given the wide geographic spread of rare earths, mining companies can make choices based on institutional and political factors as well (Asiedu and Lien 2011). Of these variables, the key factor is the strength of property rights that protect firms against the threat of expropriation.

Property rights are measured using a measure from the Varieties of Democracy project, using 2020 figures. Property rights are measured from zero to one, and capture the legal right and ability to own private property (Coppedge et al. 2021). This is an indirect measure of what concerns foreign investors – the risk of expropriation – and serves as a proxy with the assumption that countries with laws in place to protect private property specifically will also provide more protection regardless of the country of origin of the owner.[7]

A second key indicator is of rights, including "constitutionally protected civil liberties, strong rule of law, an independent judiciary, and effective checks and balances that, together, limit the exercise of executive power" and is a measure of liberal democracy (Coppedge et al. 2021). These scores come from the same source as property rights, with scores closer to one indicating more aspects of a liberal democracy and scores closer to zero fewer. People living in liberal democracies have a greater likelihood of social or environmental wrongs being publicized by a free press, and more recourse to restitution for abuses through the legal system or constitutionally protected speech and protest, while those living under illiberal regimes either lack these rights or face high risks in taking political action.

These two state-level measures are crude, but provide useful snapshots of two key measures of resilience: the ability of companies to protect their investments and feel confident about doing business and the ability of local residents to have a say in their own community's social and economic situation. Figure 3.5 shows how the 58 mining sites map onto these two dimensions.

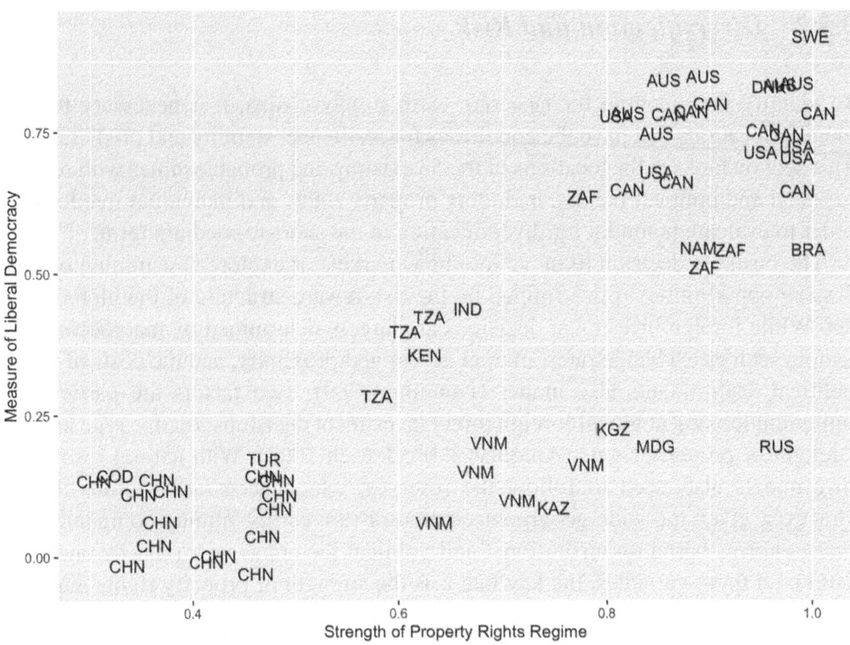

Fig. 3.5 Current and Probable Mining Sites by Institutional Status. (Notes: Regime and institution data come from Coppedge et al. (2021). Numbers closer to one represent countries with stronger property rights or stronger adherence to principles of liberal democracy. Mine data from Labay et al. (2017))

[7] In addition to this measure, I also used data from the Fragile States Index, as well as data from UNCTAD's Investor-State Dispute Settlement Database on expropriation cases to measure property rights risk. Those analyses did not substantially change the findings and are available upon request.

The sites cluster into three groups: liberal democracies with strong property rights regimes, illiberal non-democracies with strong property rights regimes, and illiberal non-democracies with weak property rights regimes. All liberal democracies in the dataset also have strong property rights protection, unsurprising given that rule of law is included in the coding definition. These sites include countries such as Australia, Brazil, Canada, Denmark (indicating the Greenland site), South Africa, Sweden, and the United States. There is a second grouping of illiberal non-democracies with stronger private property protections such as Bangladesh, Kazakhstan, or Russia. Finally, there is a group of illiberal non-democracies with weak property rights, including countries like China, North Korea, Somalia, and Tajikistan.

Figure 3.6 shows the mines from each group as percentage of total mines and a percentage of total global production in 2020. Global production percentages were calculated using data from the USGS Mineral Commodity Summary on rare earths. This institutional geography highlights risks and benefits to both mining companies and the local communities around the mines. A plurality of mines (46.5%) are in liberal democracies with strong property rights. These regimes can more firmly guarantee minimal risk of expropriation for a mining company, as well as offering *more* recourse to residents through the legal system, media, or protest movements in the event of social or environmental harm. However, only 25% of global production takes place in these countries.

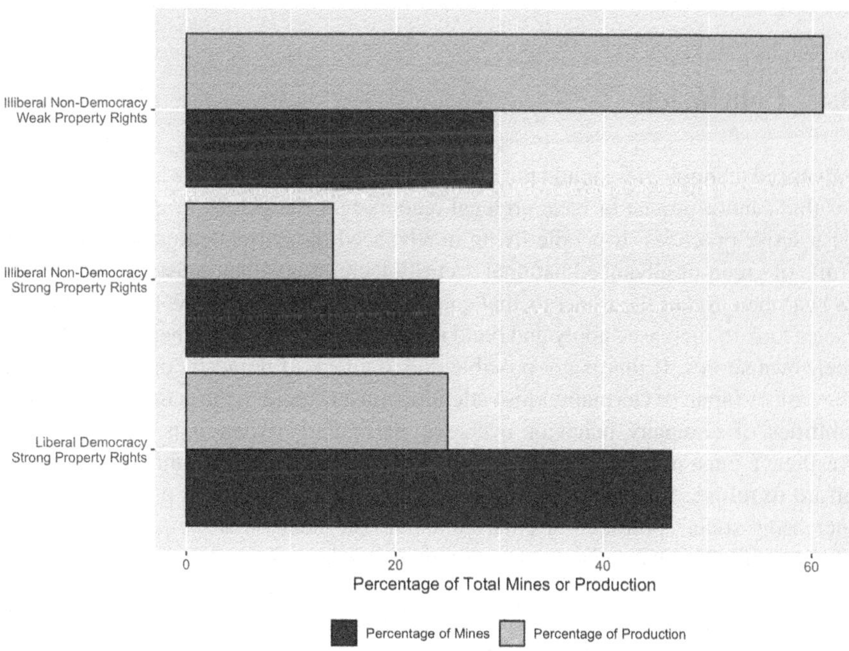

Fig. 3.6 Percentage of total mines and production by institutional status. (Notes: Data come from Coppedge et al. (2021); Labay et al. (2017); USGS (2021))

Almost a quarter of mines are located in illiberal democracies with strong property rights. Notably, these countries are places that democratic countries such as Japan has been actively seeking to open up production. For example, much of the funding from Japanese industrial policy for opening new mines or post-production sites are to countries such as Vietnam or Kazakhstan, often with limited or failed results. Only 14% of global production actually comes from this group, an indication of some of the difficulties that the rich democracies have had in pursuing international diversification arising from social, political, and environmental risks. At the same time, much of Japan's diversification drive has focused on countries within this group, and they have faced the predictable business risks often leading to failure.

Approximately a third of mines are found in illiberal non-democracies with weak property rights. China is included in this group, and accounts for 61%. Since the 1980s, expropriation threats have not been the primary political risk that concerns foreign firms. In this case, all legal rare earth mining is controlled by state-owned enterprises and inward investment in this sector is prohibited, making the threat of expropriation and foreign investment in the sector irrelevant. In order to seek non-Chinese mines and post-processing capacity and diversify the global market, countries must look to sites within the two groups with stronger property rights protection. In particular, countries should actively pursue projects in the top right quadrant of Fig. 3.5: liberal democracies with strong property rights protection and pay to internalize the risks *and* rewards of a resilient supply chain.

3.4 Conclusion

Advanced technologies should not come at the expense of people who live in societies that cannot protest or have no legal recourse for dangerous or environmentally destructive practices. If people living in advanced industrial democracies want the fruits of green or advanced national security technology, they must also be willing to bear the costs of the minerals that enable those technologies. As the rich democracies look to increase supply and build resilience, the first place they should look is their own shores. If that is not possible due to a lack of domestic deposits, such as the case of Japan or Germany, close attention must be paid to the actions and responsibilities of company behavior overseas, particularly when it is state financed. Resilience must not equate to exporting polluting practices to countries that cannot afford to refuse. Beyond the moral and ethical issue, it is also a problem because increasing social, political, or environmental risk means that the projects are also likely to fail. This has been shown in many cases throughout the world.

It is also important to acknowledge that a liberal democracy with strong property rights protection is not a panacea to all environmental or social woes. Sub-populations of these countries are still in marginalized social and political positions and may lack access to the formal court system assumed by other members of the population. Environmentally destructive mining projects in the United States have historically been in areas that are poor, remote, and inhabited by people who lack an organized political voice. The care that must be emphasized when rich countries go abroad must also be exercised within their own countries. An approach to resilience that demands those who win the benefits also bear the costs not only makes moral sense, it makes business sense as well. These lessons need to be kept in focus as the United States begins to drive towards more domestic or near-shored supply of critical minerals, for example in new investments enabled by the 2022 Inflation Reduction Act. Japan needs a similar lens with their renewed focus on critical mineral diversification in the 2022 Economic Security Legislation.

The policy history of the rare earth industry in the United States and China, as well as Japan's diversification drive, shows that a state commitment to the industry has been necessary for success. In the drive for increased resilience and supply, states must ready themselves to commit to investment in the full infrastructure of the industry, including education and training as well as mining and processing projects. States must also prepare for failures, as it is likely that will be the fate of many projects. States must maintain policy commitments to the industry such as the United States did in the postwar era, China has been doing since the 1980s, and Japan has been attempting overseas. If the rich democracies are not willing to bear the cost of domestic mining and processing themselves then trade with (and dependence on) China is not only key but is the only possible path forward.

References

13th Five-Year Plan. 2016. *The 13th Five-Year Plan for Economic and Social Development of the People's Republic of China (2016–2020).* Compilation and Translation Bureau, Central Committee of the Communist Party of China. https://en.ndrc.gov.cn/policies/202105/P020210527785800103339.pdf

14th Five-Year Plan. 2021. 第十四个五年规划和*2035*年远景目标纲要 *[China's 14th Five-Year Plan and the Long-Range Objectives Through the Year 2035 (Draft)].* National Development and Reform Commission. https://www.ndrc.gov.cn/xxgk/zcfb/ghwb/202103/P020210323538797779059.pdf

Abraham, D.S. 2015. *The elements of power: Gadgets, guns, and the struggle for a sustainable future in the rare metal age.* Yale University Press.

African Mining. 2019. REEs on a growth path. *African Mining Online*, 19 March. https://www.africanmining.co.za/2019/03/19/rees-on-a-growth-path/

Ahlquist, J.S. 2006. Economic policy, institutions, and capital flows: Portfolio and direct investment flows in developing countries. *International Studies Quarterly 50* (3): 681–704.

Ali, S.H. 2014. Social and environmental impact of the rare earth industries. *Resources 3* (1): 123–134.

Argus Media. n.d., various years. *Argus Media Rare Earth Prices.*

Asiedu, E., and D. Lien. 2011. Democracy, foreign direct investment and natural resources. *Journal of International Economics 84* (1): 99–111.

Barteková, E., and R. Kemp. 2016. National strategies for securing a stable supply of rare earths in different world regions. *Resources Policy 49*: 153–164.

Beauford, R. 2011. Mining the Moon for rare earth elements-is it really possible. *Rare Earth Elements.*

Carver, E. 2017. *Another blow to troubled Madagascar rare earth mine*, 22 November. https://news.mongabay.com/2017/11/another-blow-to-troubled-madagascar-rare-earth-mine/

Chinese Ministry of Science and Technology. 2007a. *National Basic Research Program of China (973 Program)*. https://web.archive.org/web/20070208180409/http://www.most.gov.cn/eng/programmes1/200610/t20061009_36223.htm

———. 2007b. *National High-tech R&D Program (863 Program)*. Chinese Ministry of Science and Technology. https://web.archive.org/web/20070208195727/http://www.most.gov.cn/eng/programmes1/200610/t20061009_36225.htm

Chu, D. 2010. *Seventeen metals: The middle east has oil, china has rare earth*. Business Insider, 11 November. https://www.businessinsider.com/seventeen-metals-the-middle-east-has-oil-china-has-rare-earth-2011-1

Coppedge, M., J. Gerring, S.I. Lindberg, J. Teorell, D. Altman, M. Bernhard, M.S. Fish, A. Glynn, A. Hicken, and C.H Knutsen. 2021. Varieties of democracy. *Codebook. Version*. https://www.v-dem.net/static/website/img/refs/codebookv111.pdf

Daly, T. 2021. China hikes half-year rare earth output quotas to record level. *Reuters*, 19 February. https://www.reuters.com/article/us-china-rareearth-idUSKBN2AJ18O

Dunning, J.H. 1980. Toward an eclectic theory of international production: Some empirical tests. *Journal of International Business Studies 11* (1): 9–31. https://doi.org/10.1057/palgrave.jibs.8490593.

Economy, E.C. 2011. *The river runs black*. Cornell University Press.

Fact Sheet: U.S.-Japan Competitiveness and Resilience (CoRe) Partnership. 2021. The White House, 17 April. https://www.whitehouse.gov/briefing-room/statements-releases/2021/04/16/fact-sheet-u-s-japan-competitiveness-and-resilience-core-partnership/

Gholz, E., and L. Hughes. 2021. Market structure and economic sanctions: The 2010 rare earth elements episode as a pathway case of market adjustment. *Review of International Political Economy 28* (3): 611–634.

Goldman, J.A. 2014. The US rare earth industry: Its growth and decline. *Journal of Policy History 26* (2): 139–166.

Grasso, V.B. 2013. Rare earth elements in national defense: Background, oversight issues, and options for Congress. *Congressional Research Service*. https://sgp.fas.org/crs/natsec/R41744.pdf

Haque, N., A. Hughes, S. Lim, and C. Vernon. 2014. Rare earth elements: Overview of mining, mineralogy, uses, sustainability and environmental impact. *Resources 3* (4): 614–635.

Harmer, R.E., and P.A.M. Nex. 2016. Rare earth deposits of Africa. *Episodes 39* (2): 381–406.

Henderson, P., J. Gluyas, G. Gunn, F. Wall, A. Woolley, A. Finlay, and N. Bilham. 2011. *Rare earth elements. A Briefing Note by the Geological Society of London*, 1–13. *Geological Society of London*.

Hsueh, R. 2011. *China's regulatory state*. Cornell University Press.

Humphries, M. 2019. *Critical Minerals and US Public Policy*. Washington, DC: Congressional Research Service.

Hund, K., D. La Porta, T.P. Fabregas, T. Laing, and J. Drexhage. 2020. *Minerals for climate action: The mineral intensity of the clean energy transition*. World Bank.

Hurst, C. 2010. *China's rare earth elements industry: What can the west learn?* Washington, DC: Institute for the Analysis of Global Security.

Imholte, D.D., R.T. Nguyen, A. Vedantam, M. Brown, A. Iyer, B.J. Smith, J.W. Collins, C.G. Anderson, and B. O'Kelley. 2018. An assessment of US rare earth availability for supporting US wind energy growth targets. *Energy Policy 113*: 294–305.

International Energy Agency. 2021. *The role of critical minerals in clean energy transitions*. International Energy Agency, May. https://www.iea.org/reports/the-role-of-critical-minerals-in-clean-energy-transitions

Ionic Rare Earths. 2022. *Makuutu, Uganda*. Ionic Rare Earths. https://ionicre.com.au/makuutu-uganda/

ITALaw. 2020. *Cortec Mining Kenya Limited, Cortec (Pty) Limited and Stirling Capital Limited v. Republic of Kenya, ICSID Case No. ARB/15/29 | italaw*. https://www.italaw.com/cases/3974

Jensen, N.M. 2003. Democratic governance and multinational corporations: Political regimes and inflows of foreign direct investment. *International Organization 57* (3): 587–616.

JETRO. 2020. 第一回公募(設備導入補助型(一般枠・特別枠))における採択事業者について | 海外サプライチェーン多元化等支援事業のサービス. ジェトロ, 17 July. https://www.jetro.go.jp/services/supplychain/kekka-1.html

JOGMEC. 2020. *JOGMEC Integrated Report 2020*. http://www.jogmec.go.jp/content/300373866.pdf

———. 2022. *Financial support for Japanese companies: Metals*. Japan Oil, Gas and Metals National Corporation (JOGMEC). https://www.jogmec.go.jp/english/stockpiling/metal_10_000003.html

Johnson, C. 1982. *MITI and the Japanese miracle: The growth of industrial policy, 1925–1975*. Stanford University Press.

Kennedy, J. 2019. China solidifies dominance in rare earth processing. *National Defense Magazine*.

Kilby, C. 2014. China's rare earth trade: Health and the environment. *The China Quarterly 218*: 540–550.

Klinger, J.M. 2018. *Rare earth frontiers: From terrestrial subsoils to lunar landscapes*. Cornell University Press.

Labay, K. M.H. Burger, J.D. Bellora, K.J. Schulz, J.H. DeYoung, R.R. Seal II, J.L. Mauk, D.C. Bradley, and C.A. San Juan. 2017. *Global distribution of selected mines, deposits, and districts of critical minerals*.

Li, J. 2012. 民族经济可持续发现战略研究 *[Minzu Jingji Ke Chixu Faxian Zhanlue Yanjiu; Study of Sustainable National Economic Growth Strategies]*. Zhongguo Zaizhengjingji Publishing.

Lynch, D. 2019. China hints it will choke off U.S. 'rare earths' access. But it's not that easy. *Washington Post*, 10 June. https://www.washingtonpost.com/business/economy/2019/06/07/80a06794-8649-11e9-a491-25df61c78dc4_story.html

Maehara, S. 2011. *Foreign policy speech by minister for foreign affairs Seiji Maehara to the 177th Session of the Diet*, 24 January. https://www.mofa.go.jp/announce/fm/maehara/speech110124.html

Mancheri, N., and T. Marukawa. 2016. *Rare earth elements China and Japan in industry, trade and value chain*.

Mancheri, N.A., B. Sprecher, G. Bailey, J. Ge, and A. Tukker. 2019. Effect of Chinese policies on rare earth supply chain resilience. *Resources, Conservation and Recycling 142*: 101–112.

minedat.org. 2019. *Mountain Pass Mine, Mountain Pass, Mountain Pass District, Clark Mts (Clark Mountain Range), San Bernardino Co., California, USA*. https://www.mindat.org/loc-11616.html

Ministry of Economy, Trade, and Industry. 2010a. *Heisei 22 Nendo Rea Aasu nado Riyou Sangyou nado setsubi dounyuu hojokin (ichiji koubo)—Saitaku jigyou ichiran. [Industries and Facilities that Use Rare Earths (Public Appeal) – List of Selected Companies]*.

———. 2010b. *Rea Aasu nado riyou sangyou nado setsubi dounyuu jigyou—Koubo youryou [Industries and Facilities that Use Rare Earths – Public Appeal]* [Policy Document].

———. 2021. *11th conference on critical materials and minerals held*, 30 June. https://www.meti.go.jp/english/press/2021/0630_002.html

Nassar, N.T., J. Brainard, A. Gulley, R. Manley, G. Matos, G. Lederer, L.R. Bird, D. Pineault, E. Alonso, and J. Gambogi. 2020. Evaluating the mineral commodity supply risk of the US manufacturing sector. *Science Advances 6* (8): eaay8647.

Obulutsa, G., and H. Malalo. 2018. Kenya wins case against Cortec Mining on rare earths, niobium licence. *Reuters*, 23 October. https://www.reuters.com/article/kenya-mining-idUSL8N1X335S

Pempel, T.J. 2021. *A region of Regimes: Prosperity and Plunder in the Asia-Pacific*. Cornell University Press.

Rainbow Rare Earths. 2022. Gakara project. *Rainbow Rare Earths*. https://www.rainbowrareearths.com/project/gakara/

Reuters. 2019. China to step up crackdown on rare earth sector: Ministry. *Reuters*, 4 January. https://www.reuters.com/article/us-china-rareearths/china-to-step-up-crackdown-on-rare-earth-sector-ministry-idUSKCN1OY0R3

Schulz, K.J., J.H. DeYoung Jr., R.R. Seal II, and D.C. Bradley. 2017. *Critical mineral resources of the United States—An introduction*, Report No. 1802A; Professional Paper, p. 22. USGS Publications Warehouse. https://doi.org/10.3133/pp1802A.

Solis, M. 2004. *Banking on multinationals: Public credit and the export of Japanese sunset industries*. Stanford University Press.

———. 2021. The big squeeze: Japanese supply chains and great power competition. *Joint U.S.-Korea Academic Studies*.

Standaert, M. 2019. China wrestles with the toxic aftermath of rare earth mining. *Yale E360*.

Sukumaran, T. 2019. Malaysia's decision to renew rare earth miner Lynas' operating licence could come back to bite ruling coalition. *South China Morning Post*, 1 September. https://www.scmp.com/week-asia/economics/article/3025151/malaysias-decision-renew-rare-earth-miner-lynas-operating

Toro, N., P. Robles, and R.I. Jeldres. 2020. Seabed mineral resources, an alternative for the future of renewable energy: A critical review. *Ore Geology Reviews* 126: 103699.

USGS. 2021. *Mineral commodity summaries*. United States Geological Survey.

Vekasi, K. 2019. Politics, markets, and rare commodities: Responses to Chinese rare earth policy. *Japanese Journal of Political Science 20* (1): 2–20.

Vogel, S.K. 2018. *Marketcraft: How governments make markets work*. Oxford University Press.

Woo-Cumings, M. 1999. *The developmental state*. Cornell University Press.

World Trade Organization. 2015. *DS431: China—Measures Related to the Exportation of Rare Earths, Tungsten and Molybdenum*, 20 May. https://www.wto.org/english/tratop_e/dispu_e/cases_e/ds431_e.htm

Zhou, Q., and S. Brooke. 2022. China merges three rare earths state-owned entities to increase pricing power and efficiency. *China Briefing News*, 12 January. https://www.china-briefing.com/news/china-merges-three-rare-earths-state-owned-entities-to-increase-pricing-power-and-efficiency/

Kristin Vekasi is an Associate Professor in the Department of Political Science and School of Policy and International Affairs at the University of Maine. Her research focuses on trade and investment strategies in changing geopolitical environments, and the political risk management of supply chains. She specializes in Northeast Asia, and has spent years conducting research in China, Japan, and South Korea. Her book Risk Management Strategies of Japanese Companies in China (Routledge 2019) explores how Japanese multinational corporations mitigate political risk in China. Her current research examines how Japan, China, and the United States cooperate and compete to manage complex supply chains in Southeast Asia, focusing on industries essential for the transition to green energy. Vekasi received her PhD in political science from the University of Wisconsin, Madison. Prior to joining the faculty at University of Maine, she taught at New College of Florida, was a visiting Research Fellow at the University of Tokyo and a Fulbright Fellow at Tohoku University. She is a member of the Mansfield Foundation's US-Japan Network for the Future, and a 2019 National Asia Research Program Fellow with the National Bureau of Asian Research where she is also a nonresident fellow. In 2021–2022, she was an academic associate at the Harvard University US-Japan Program.

Part II
Critical Minerals and their Supply Chains

Chapter 4
Public Policy Toward Critical Materials: A False Dichotomy, a Messy Middle Ground, and Seven Guiding Principles

Roderick G. Eggert

Abstract What kind of public policies should the United States and other market economies have toward critical materials? A false dichotomy often exists in discussions about national strategies, organized around two caricature-like, mutually exclusive approaches to policy. First, let the market decide, limiting government activities to establishing legal and regulatory frameworks for efficient and effective decision making about private initiatives and activities (a sector-agnostic approach). Alternatively, have governments pick winning (and, implicitly, losing) commercial sectors, companies, technologies and products – sometimes referred to simplistically as the 'industrial policy' approach. In reality, most governments can, should, and do operate in the messier middle ground between these two extremes. This chapter discusses this messy middle ground and proposes seven principles to guide development of public policy toward critical materials.

Keywords Critical materials · Industrial policy · Supply chain risks · Rare earths · Decarbonization

4.1 Introduction

Critical materials provide essential properties to engineered materials and systems. Selected rare-earth elements, for example, are essential in state-of-the-art permanent magnets and, in turn, technology at the center of decarbonization – namely the electric motors for electric vehicles, wind turbines and many household products such as air conditioners. Rare-earth magnets also are essential in some military technologies. Other critical materials include those used in batteries and energy storage, such as lithium, cobalt, nickel and vanadium; and electronic materials in solar technologies and power electronics, such as gallium, indium and tellurium.

R. G. Eggert (✉)
Department of Economics and Business, Colorado School of Mines, Golden, CO, USA
e-mail: reggert@mines.edu

S. Kalantzakos (ed.), *Critical Minerals, the Climate Crisis and the Tech Imperium*, Archimedes 65, https://doi.org/10.1007/978-3-031-25577-9_4

In addition to being vital or indispensable, critical materials – as the term 'critical' has come to be used – have a second characteristic: They are subject to supply-chain risks. Some of these risks are short term, usually reflecting lack of geographic or company diversity in one or more stages in current supply chains, often with geopolitical risks. Rare earths fall into this category because of the geographic concentration of production in China and the concern that Chinese producers or the Chinese government will act opportunistically to the disadvantage of customers outside of China. Sometimes the short-term risk is associated with the small, opaque markets typical of most specialty metals; lack of transparency makes these markets subject to unexpected disruption.

Other risks are longer term, reflecting the fear that the supply of materials, relative to growing societal demand, will not be "appropriate" – the concern that supplies will not be sufficient, affordable, secure, sustainable (in an environmental sense), and responsible (in a social sense). Lithium, essential in lithium-based batteries – provides an example of this longer-term risk – namely that lithium production will not increase fast enough, at affordable prices, and in ways that are sufficiently sustainable and responsible to meet the requirements of electric vehicles and grid-scale storage of energy from wind and solar power.

Beyond defining a critical material, a bit of history is important in setting the stage. Over the last half century, supply chains have become more dispersed, fragmented and (often) concentrated geographically, as globalization has allowed each stage of a supply chain to be located where costs are lowest (Tilton and Guzmán 2016).[1] But low costs, in some cases, are associated with supply-chain risks due to geopolitics, potential for natural disasters, blockages of key transportation routes and other factors.

As a result, many national governments are asking, what kind of policy should they pursue – for consuming nations, how to manage supply-chain risks; for producing countries, how to help consuming nations mitigate their risks and simultaneously expand and enhance mining and industrial activities in their countries? This chapter focuses on the perspective of countries that are primarily users of critical materials, such as the United States, Japan, and the European nations. It does not offer specific policy recommendations but rather provides a framework in which to consider public policy toward rare earths and other critical materials – including seven guiding principles.

[1] However, globalization has not been without its critics. Lynn (2005) warned of what he referred to as "the rise and the coming fall of the global corporation from the backlash against globalization.

4.2 A False Dichotomy

How a nation that is an important user of materials manages its supply-chain risks fundamentally reflects choices it makes about industrial strategy. A false dichotomy often exists in discussions of public policy, organized around two almost caricature-like, mutually exclusive approaches to public policy and industrial strategy.

First, *let the market decide*: Let the market decide which industries to emphasize and in what ways, limiting government activities to establishing the legal and regulatory frameworks for efficient and effective decision making about private initiatives and activities – a sector agnostic approach. This approach emphasizes well-defined property rights and minimal but efficient regulation of commercial activities. If barriers to international trade create supply-chain risks, then government should work through multilateral institutions – such as the World Trade Organization – to remove these trade barriers. If inefficient approval processes for permitting new mines inhibit development of mines that could help mitigate supply-chain risks, then improve these permitting processes. If (private-sector) manufacturers face supply-chain risks for raw materials, they are in the best position to manage these risks themselves according to this perspective. National defense is an exception; but even here, think of a nation's military as a customer facing supply-chain risks. The military should evaluate and manage its supply-chain risks as if it were a private-sector user of materials but using national security rather than profits as the primary goal.

Second, *pick winners and losers*: In this approach, governments pick winning (and, implicitly, losing) commercial sectors, companies, technologies and products – sometimes referred to as 'industrial policy.' This approach uses policy tools – such as restrictions on international trade (tariffs and quotas), and tax incentives and subsidies – to favor domestic companies and sectors deemed essential and to assist the commercial sector in managing supply chain risks.

In reality, most governments operate in the messier middle ground between these two extremes – asking the question, what is the right mix of market solutions and government direction toward commercial activities and technology development?

4.3 The Messy Middle Ground

The middle ground is more nuanced and less "either/or" than the caricatures above suggest. In fact, market and industrial-policy solutions to supply-chain risks – rather than being mutually exclusive – are complementary. There are four dimensions of this middle ground.

First, *a more-nuanced approach to "let the market decide."* Economists recognize that markets often do not, by themselves, correct for the distortions of externalities, adequately provide public goods, and promote equity or fairness. It is possible to rely primarily on market solutions to problems of supply-chain risks while having government intervene to correct for these market failures. To be sure, economists disagree significantly about the nature and degree of government intervention that is appropriate in these circumstances. In the case of externalities, an externality is a spillover effect on entities not involved in making a decision. The spillover effect may be positive or negative. In mineral production and material manufacturing, the damages caused by pollution are examples of negative externalities. Government policy plays an important role in ensuring that the negative externalities of pollution are incorporated into decision making about activities that create the pollution.

Moving from externalities to public goods, public goods are underprovided from society's perspective by private decision making alone because it is difficult for private providers of these goods to fully capture the benefits of investing in these goods (Mankiw 2020). More formally, public goods are nonrival, meaning that my use of the good does not reduce your ability to also use or benefit from a good. Think of a lighthouse that warns all ships of proximity to a coastline. A public good also is nonexcludable, meaning that it is difficult to exclude someone from benefiting from a good – even if a person does not pay for the good. Think of knowledge created through basic research. For one or a combination of both reasons, public goods will be underprovided from the perspective of society as a whole. In practice, some goods are partly private and partly public, such as education and applied research and development.[2] Public policy toward education and research are so important to the discussion of critical materials that I discuss it later in this essay as its own dimension of policy toward critical materials.

As for equity, this concept as used here represents fairness in the distribution of benefits and burdens associated with an activity. In mineral production and material manufacturing, the burden of pollution and the social disruption of a boom-and-bust industry often are borne by local communities. Yet the benefits often accrue to national governments and private investors outside the local community. This allocation of benefits and burdens often is deemed unfair or inequitable. A government might intervene to re-allocate the benefits and burdens of mineral production and manufacturing in favor of local communities. Increasingly, nongovernmental actors – including private companies – are involved in these issues in what has come to be called environmental, social and governance (ESG) considerations about how companies perform beyond earning profits (see, for example, Business Roundtable 2019; Kaplan and Ramanna 2021).

[2]Another way to think about public goods is that they are not private goods. Think of a loaf of bread as a private good. A loaf of bread is rival in the sense that, if I eat it, you cannot also eat it. It is excludable in that a store can make it difficult for me to acquire a loaf of bread if I do not pay for it. For private goods, there is greater alignment between incentives facing private providers of the good and societal goals than is true for public goods. In other words, what is optimal for private providers and what is optimal for society as a whole are more likely the same for private goods than for public goods. See any microeconomics textbook, such as Mankiw (2020).

Second, *public policy toward education, science, technology and innovation.*[3] As noted above, education and research are at least partly public goods. As such, governments can play an important role in facilitating the provision of these goods. The starting point for thinking about education and research is to be largely sector agnostic and prioritize education and basic research – that is research aimed primarily at creating new knowledge rather than to solve a particular societal problem. The priority is the longer term, a decade or more into the future. The focus of public policy is providing the inputs that private organizations and entrepreneurs need to be productive, innovative and successful, namely educated people and knowledge developed through basic (pre-competitive) research. Think of education and pre-competitive research as forms of knowledge infrastructure that government, together with the private sector, need to provide. Public funding anchors education and basic research.

An extension of this emphasis on education and basic research is to prioritize strategic sectors or technologies. Public policy is no longer sector agnostic but rather targets strategic sectors or technologies – for example, batteries or electric motors. Policy does not, however, favor specific types of technology within a technology area – for example lithium-ion batteries – or specific companies but encourages competition among technologies and companies. Public funding anchors these activities.

For raw-material supply chains, this approach might focus on education and research in geoscience and economic geology, mineral processing and extractive metallurgy, materials science and engineering, advanced manufacturing, recycling and reuse – as well as broad, scene-setting strategic analysis. The focus would be education and research aimed at diversifying primary production; wasting less of what we do produce through improvements to manufacturing efficiency, material re-use and recycling and reuse; and developing substitute materials to reduce dependence on critical raw materials. The result would be technological options available for the private sector to test, develop, and potentially commercialize – ultimately reducing supply-chain risks.

Public policy might aim to develop industry-led collaborations for priority raw-material supply chains, perhaps modeled after Sematech and semiconductors in the United States in the 1980s and 1990s.

A further extension is to include publicly funded efforts focused on bridging the gap between basic research and full deployment of new technologies. This extension has shorter time horizons than the earlier focus on basic research; it focuses on the short to medium term (one or a few years, up to about a decade). It takes ideas and knowledge generated in basic, laboratory-scale research and prioritizes accelerated testing of them in larger pilot and demonstration facilities. It focuses on evaluating whether a good idea in the laboratory can be scaled and deployed commercially. Public-private partnerships, with funding from both government and the private

[3] Motivated by and adapted from David Sainsbury's approach to innovation policy in his book *Windows of Opportunity: How Nations Create Wealth* (2020).

sector, are a key organizational form that reflects the partly public, partly private nature of investment in scale-up activities. Stated slightly differently, government activities focus on accelerating the pace of converting new knowledge and specific commercial goals into commercial activities in a more targeted way.

In this second dimension of the messy middle ground, public policy focuses on innovation as a key to mitigating and managing supply-chain risks. Innovation, of course, can be and usually is copied by competitors and to some degree benefits everyone, even those companies and countries that do not pay for the research and development leading to the innovation. There is free riding, in other words, on the efforts of others. Nevertheless, the rationale for innovation as a key to managing supply-chain risks relies on one or a combination of two arguments: (a) focus innovative activities on processes or products that benefit the home country but not other countries or companies that are sources of supply-chain risk, and (b) worry less about the potential that a specific innovation is copied and focus, instead, on innovating faster than competitor nations and keeping one step ahead even if innovations are copied.

Third, *a more-nuanced commercial policy of "picking winners and losers" to achieve results sooner than relying on purely market solutions, as well as to prioritize national objectives that are insufficiently considered by private users of materials.* This approach prioritizes having commercial impact in the short to medium term in the belief that market solutions take too long to materialize. Focus is on steering market activities actively, deliberately and quickly toward specific commercial sectors and technology areas, such that supply-chain risks are mitigated.

A part of this argument is that private companies, especially those with operations in many countries, will not give sufficient priority to narrower national priorities. For example, a company with operations in both China and the United States will evaluate risks differently than might be optimal for the United States alone.

Among possible policy actions are:

- Create and manage economic stockpiles of critical raw materials, in cooperation with industry. Government and the private sector together fund and manage working inventories of materials deemed essential – not materials for defense purposes but key inputs for the private manufacturing sector, such as steel, automobiles, electronics and others. Government arguably might help the private sector manage its supply-chain risks if government had better insights into risks than individual companies or if there are economies of scale in acquiring materials for and in managing working stockpiles. Japan and South Korea have such economic stockpiles of materials (for Japan, see Japan Organization for Metals and Energy Security 2022).
- Use loan guarantees open to all companies to attract financing for risky investments that have the potential to mitigate supply-chain risks.
- Fund or co-fund investments outside the domestic economy that would mitigate supply-chain risks. The Japan Organization for Metals and Energy Security makes such investment on behalf of the Japanese economy.

– Use subsidies or tax incentives to encourage domestic investment in activities that would mitigate supply-chain risks.

The big danger in this approach is government capture – namely that private interests with the most effective lobbying presence steer public funds toward their own activities regardless of the merits of devoting public funds to these activities. In other words, public funding supports activities that have largely private, rather than more broadly public, benefits.

Fourth, *international collaborations and trade policy.* A focus on international collaborations recognizes that rarely does one nation have a competitive advantage in all stages of a supply chain. This approach emphasizes working with allied nations to manage supply-chain risks jointly; an example is the Minerals Security Partnership established in 2022 by the United States and a number of other nations including Australia, Canada, Japan and the European Union (International Energy Agency 2022) with the goals of strengthening information sharing, increasing investment in supply chains for critical minerals and developing recycling technologies. More generally, among possible international initiatives are government-to-government collaborations to:

– Assess resource potential in unconventional primary resources,
– Assess recovery potential of valuable materials from process waste streams of both historical and ongoing mining and metallurgical operations – giving priority to specialty metals present in dilute concentrations, often not recovered and when recovered are minor byproducts of the principal products of an operation,
– Assess potential for circularizing material life cycles,
– Develop economic stockpiles across multiple nations,
– Develop and articulate specific strategic goals for re-shoring or near-shoring entire supply chains for selected minerals and materials necessary for clean energy (e.g., rare earths, magnets, motors) or for other areas of strategic priority,
– Develop public-private partnerships and joint funding to finance demonstration and pilot facilities aimed at technological de-risking and accelerating the deployment of new technologies, and
– Provide public funding to partially support investment in standby production capacity for specialty metals (e.g., gallium, germanium, indium) that often reside in mining and metallurgical waste streams.

These joint actions have time horizons ranging from short term to long term. Of course, activities in all of these areas also could be pursued at a narrow national level.

Policy toward international trade starts with seeking multilateral reductions in barriers to trade, to better allow specific stages in a supply chain to be located where costs are lowest in the absence of trade barriers. Trade policy should not include imposing trade restrictions on all foreign sources of critical materials, even if import dependence is high. Rather if import dependence is deemed an important source of supply-chain risk – and it does not need to be if imports come from a diverse set of exporting nations or companies – then focus trade policy on these risky sources. But recognize that a significant problem with limiting trade is that trade restrictions act like a tax on domestic customers of a material. It is better to focus public policy directly on establishing a supply chain independent of the risky parts of the supply chain.

Any discussion of trade policy and critical materials needs to consider China and the large market shares of production in China for a number of critical materials. Some have called for U.S. decoupling from China (see, for example, Navarro and Autry 2011, and many public statements from officials in both the Trump and Biden administrations). Others caution that full decoupling is illusory and potentially very damaging to the U.S. economy, just as it would be damaging to the Chinese economy (see, for example, Farrell and Newman 2020; U.S. Chamber of Commerce 2021). A complete discussion of U.S. public policy toward China is outside the scope of this paper.

4.4 Seven Guiding Principles

Critical materials are diverse. Nearly every chemical element in the periodic table of elements is vital or essential to some technology (e.g., electric vehicles), company (e.g., Nucor steel or General Electric), household (e.g., mine) or other customer (e.g., U.S. Department of Defense) – and which materials are indispensable will vary from case to case. Moreover, each critical material has its own story in terms of the sources of supply-chain risks. In the short term, as noted earlier, risks are due primarily to lack of diversity in existing supply chains, especially when geopolitical risks are present. In the long term, the major concerns are that future supplies may not increase as quickly as manufacturers might desire (to, for example, allow for desired deployment of electric vehicles); that prices may be higher than customers are willing to pay for some priority uses (such as for battery materials in electric vehicles); that production may be more energy intensive, environmentally damaging, or socially disruptive than we like; and that lack of innovation and paucity of skilled workers will stymy solving critical material problems. Which concerns are paramount, whether short term or long term, vary from material to material. Finally, what is 'critical' is dynamic. Materials that are critical today may not be critical in the future as today's supply-chain problems are solved and as end-use technologies – and their material inputs – change.

Given this diversity of material uses and sources of risk, as well as the dynamic nature of what is 'critical,' what should governments do? My personal bias as an economist is to anchor public policy toward critical materials around market solutions – but market solutions and public policies that start with, but are not limited to, the caricature I provided at the beginning of this essay; they should, very selectively, incorporate aspects of industrial policy. More specifically, I suggest the following, seven guiding principles when thinking about the messy middle ground between market solutions and industrial policy:

1. Rely on markets but recognize the time lags in market solutions to critical-materials challenges,
2. Emphasize government's role in facilitating long-term solutions, especially through education and basic research in areas over the entire life cycle of a critical material,
3. Pursue innovation policy that targets strategic technology areas without favoring specific technologies or companies within the area,
4. Recognize the important role government can play in linking basic research (a public good) and commercial processes and products (private goods) through industry-inspired applied research (which is partially private and partially public) and co-funded pilot and demonstration facilities,
5. (Very) selectively use other tools of industrial policy – such as loans, loan guarantees, and government co-financing of commercial investment – in circumstances in which there clearly is a national strategic interest in fostering commercial activities quicker than would occur by relying on markets by themselves,
6. Generally avoid international-trade restrictions, and
7. Avoid focusing public policy around a goal of national self-sufficiency. Different stages of a critical-material supply chain should be located where costs are lowest even if different stages are in different countries – as long the geography of a supply chain has acceptable risks. Use international collaborations to foster supply chains that are diverse and low cost.

Acknowledgement and Disclaimer This work is supported in part by the Critical Materials Institute, an Energy Innovation Hub funded by the U.S. Department of Energy, Office of Energy Efficiency and Renewable Energy, Advanced Materials and Manufacturing Technologies Office. The views expressed in this chapter, nevertheless, are mine alone and do not represent the official views of the U.S. Department of Energy.

Note The starting point for this chapter was written testimony I prepared for a hearing of the Canadian House of Commons' Standing Committee on Natural Resources and its study of critical minerals and associated value chains in Canada, March 26, 2021. A second version benefitted from discussions at the virtual workshop Rich Rocks, the Climate Crisis, and the Tech Imperium, July 12–14, 2021, organized by the Research Institute for the History of Science and Technology at Caltech and The Huntington.

References

Business Roundtable. 2019. *Statement on the purpose of a corporation: Delivering long-term value to all stakeholders*, 19 August. https://www.businessroundtable.org/purposeanniversary. Accessed 23 Jan 2022.

Farrell, H., and A. Newman. 2020. The folly of decoupling from China: It isn't just Perilous – It's impossible. *Foreign Affairs*, online 3 June 2020. https://www.foreignaffairs.com/articles/china/2020-06-03/folly-decoupling-china?utm_source=google&utm_medium=cpc&utm_campaign=gap_ds&gclid=Cj0KCQiAuP-OBhDqARIsAD4XHpf7HgiaxtFHUEvFs2z9Z41Ksk1I1nEvLfru8SwX-TjY2Bf6tf7sW3YaArpJEALw_wcB. Accessed 13 Jan 2022.

International Energy Agency. 2022. "Minerals Security Partnership," https://www.iea.org/policies/16066-minerals-security-partnership. Accessed 25 Nov 2022.
Japan Organization for Metals and Energy Security. 2022. Website: https://www.jogmec.go.jp/english/. Accessed 25 Nov 2022.
Kaplan, R.S., and K. Ramanna. 2021. Accounting for climate change: The first rigorous approach to ESG reporting. *Harvard Business Review*, November–December 2021 issue.
Lynn, B.C. 2005. *End of the line: The rise and coming fall of the global corporation*. New York: Doubleday.
Mankiw, N.G. 2020. *Principles of microeconomics*. 9th ed. Boston: Cengage Learning.
Navarro, P.W., and G. Autry. 2011. *Death by China: Confronting the Dragon – A global call to action*. Upper Saddle River/New Jersey: Pearson FT Press.
Sainsbury, D. 2020. *Windows of opportunity: How nations create wealth*. London: Profile Books.
Tilton, J.E., and J.G. Guzmán. 2016. *Mineral economics and policy*. New York: Routledge.
U.S. Chamber of Commerce. 2021. *Understanding U.S.-China decoupling: Macro trends and industry impacts*. https://www.uschamber.com/assets/archived/images/024001_us_china_decoupling_report_fin.pdf. Accessed 13 Jan 2022.

Roderick G. Eggert is Viola Vestal Coulter Foundation Chair in Mineral Economics at Colorado School of Mines, where he has taught since 1986. He also is Deputy Director of the Critical Materials Institute, a research consortium established by the U.S. Department of Energy in 2013 to accelerate innovation in energy materials. He chairs the Independent Advisory Board of the UKRI Circular Economy Centre for Technology Metals (United Kingdom) and serves on the Advisory Board of the international Rare Earth Industry Association. His research and teaching focus on mineral economics and public policy. He chaired the U.S. National Research Council committee that wrote the 2008 book *Minerals, Critical Minerals, and the US Economy* (National Academies Press). He has testified on mineral issues to committees of the U.S. Senate and House of Representatives, the Canadian House of Commons and European Parliament. He has a B.A. in earth sciences from Dartmouth College, a M.S. in geochemistry and mineralogy from Penn State University, and a Ph.D. in mineral economics also from Penn State.

Chapter 5
Lessons from Three Decades in the Rare Earth Trenches

Constantine Karayannopoulos and Vasileios Tsianos

Abstract Over the past several decades, global rare earth markets have ridden a dizzying roller coaster that has, at times, shaken the faith of industrial consumers in the reliability of supply chains for these critical materials. Today, rare earth production and associated supply chains are stronger and more diversified than at any other time in the last two decades. The industry, however, faces increasing pressure to expand to new levels of production and downstream sophistication in order to meet rising demand driven by advanced technologies and global decarbonization. As governments work to encourage more rare earth production around the world and, in particular critical cases, craft industrial strategies to ensure resilient operations of rare earth supply chains, it is important to understand the evolution of this industry. This chapter examines relevant events and interconnected factors – economic, technological, and geopolitical – that fashioned the current rare earth industrial landscape. It seeks to demystify and correct some misperceptions that persist. Finally, it makes reference to industrial and policy strategies that are being implemented with some success to encourage the continued growth and resiliency of this innovative and enormously important industry and draw lessons for the successful establishment of resilient, diversified, vertically integrated rare earth supply chains.

Keywords Permanent magnets · Neo performance materials · Magnequench · Raw materials act · Neodymium · Praseodymium

5.1 Introduction: Economic First Principles

The contemporary perception of the rare earth industry's current geographic concentration in Asia is that it evolved because of the intersection of geopolitical forces and environmental policies. In other words, the People's Republic of China (PRC)

C. Karayannopoulos · V. Tsianos (✉)
Neo Performance Materials, Toronto, ON, Canada
e-mail: c.karayannopoulos@neomaterials.com; v.tsianos@neomaterials.com

S. Kalantzakos (ed.), *Critical Minerals, the Climate Crisis and the Tech Imperium*, Archimedes 65, https://doi.org/10.1007/978-3-031-25577-9_5

developed its global market share dominance of rare earth supply chains as a means of achieving larger geopolitical goals. At the same time, the West effectively hindered the development of these supply chains through burdensome environmental regulations that made rare earth mining and processing uneconomic vis-à-vis production in China, where, according to popular but sadly uninformed western lore, producers operate free of environmental constraints.

After spending almost 30 years in this industry, I argue a different thesis: fundamental demand-driven industrial economics, coupled with a lack of coherent industrial policies in the West, led to the current supply chain map. It is the high growth of downstream industry participants and Original Equipment Manufacturing ("OEM") customers, either inside or in the proximity of the PRC, that have driven the rare earth industry's growth in China, not China's desire to dominate global rare earth markets as an end in itself.

While there is a century-long industrial history of rare earths, in this chapter I will focus mainly on the last 35 years, which coincide both with the majority of my career in this industry and with several tectonic shifts that dominated the evolution of rare earth supply chains during this period. These tectonic shifts were a natural consequence of basic economics and national industrial policies, or the lack thereof. As executives in the industry for three decades, my colleagues and I have had first-row seats to the decisions that businesses and governments made directly affecting this industry over the years. In some cases, those decisions facilitated adaptation to the changing landscape. In others, they led to policy and business failures.

In my view, one major lesson to be learned from the evolution of rare earth supply chains is this: as they say in real estate, it is all about location, location, location. In the case of rare earths, that has meant that rare earth supply chains tend to develop in geographic proximity to their downstream customers and value-added manufacturing facilities, not necessarily near rare earth mining and processing facilities. Why is this the case?

Supply chains are eminently rational. They form efficiently – over time – in response to market pull, and they build on available critical skills. In many ways, successful supply chains are a virtuous cycle of demand, skills development, value creation, increasing investment, and ever-increasing efficiencies. Market pull forces an inevitable behavior through the supply chain. That, in turn, creates skill sets that didn't exist before. These skill sets enhance the competitive capacity of the supply chain and deliver value to all segments involved. As customers pull suppliers, it forces suppliers to innovate, adapt business models, and increase operational efficiencies. Eventually, in a successful supply chain, everyone achieves appropriate returns on capital as well as sustained positive cash flows.

The opposite is also true: if a link in the chain is not profitable, it becomes the weak link; it breaks; and the entire supply chain suffers. The critical point I want to convey here is this: all segments of the virtuous cycle are driven by market demand/customer pull, not by supply/production push. The theory that "if you build it, they will come" is a compelling narrative for Hollywood movies. In reality, supply chains tend to form around the Tesla companies of the world, not the upstream mines that feed into Tesla's batteries and electric vehicles. Thus, as the world's manufacturing infrastructure has become increasingly global and interconnected, it is not uncommon today to see the rare earth and other critical material supply chains that feed a single automotive manufacturer, traversing more than a dozen nations and supply chain steps from the mine to the finished vehicle.

This is one reason why the increasing calls by policymakers in various nations to decouple their economy from supplies of imported critical minerals may not meet with quick success in the real world. If Nation A seeks "independence" from the rare earths mined in Nation B, and the leaders of Nation A invest the political and economic capital (and years) necessary to establish domestic rare earth mining and processing, the supply chain that leads from those domestic mines to their domestic manufacturing of refrigerators or wind turbines or electric vehicles will almost certainly still traverse multiple nations.

To expect that, in capitalist systems, rare earth mine-to-vehicle supply chains would naturally develop in only one geography flies in the face of the fundamental tenets of comparative advantage and rational economic behavior. Left to its own devices, free markets do not operate independently of their natural proclivities toward efficient resource allocation, continuous productivity and efficiency improvements, and the pursuit of a maximum return to investors and stakeholders, independent of geography.

Sometimes, governments need to step in with smart policies and even regulations. Larry Fink, BlackRock's CEO, eloquently made the point at a Morgan Stanley ESG conference a while back: *"The role of government is to save capitalism from itself"* (Fink 2021). One could argue that the Chinese took a portion of this maxim to heart as they built their modern manufacturing powerhouse. China's development model, of course, is somewhat unique, with its mix of long-horizon planning, state-directed "industrial policy," and mechanisms to stimulate or influence consumer demand. While not appropriate for Western democracies, this model allowed the PRC to develop a large manufacturing base and fully integrated critical mineral supply chains (frequently starting with raw materials originating very far afield with mines in Africa, Australia, and the Americas) to feed such manufacturing. China encouraged utilization of value-added rare earth materials in finished goods and consumer products that were to be "Made in China."

It should not come as a surprise that, while China may be the world's dominant producer of rare earths, it also is the world's dominant consumer of rare earths. In fact, I would argue that it is the largest producer of rare earths because it is the largest consumer of rare earths. This industrial policy is also why China's engine of technological innovation continues running strong. Electric vehicles are a good case in point.

As students of history, China's leaders understand that automotive production has been an indispensable pillar of virtually every single post-WWII industrial economy. They know that you cannot have a robust and modern industrial economy without a robust automotive industry or a number of national corporate champions critical to every link in the automotive supply chain. To enable such an industry, PRC policymakers focused on developing China's rare earth supply chain as a means of enabling and securing the systems used in electrified vehicles: motors, batteries, drivetrains, electronics, etc. They directed public and incentivized private investment and allocated capital towards all segments of their automotive supply chains, and E.V. supply chains in particular. Most important, they employed a combination of incentives and mandates to stimulate consumer demand from their rapidly growing middle class for these platforms (Interesse 2022). Tesla experienced China's success in this effort when it went to China and found already mature and established supply chains for the production of its E.V.s. Today, Tesla E.V.s are being built in China at a higher quality and lower cost according to relevant reports. Ultimately, the power of consumer market pull for E.V.s and the corresponding response by the supply chains is why China now leads the world in electric passenger cars – both manufacturing and sales – with 1.2 million vehicles sold in 2020. That dwarfs the second and third largest markets of the European Union (745,684) and the U.S. (328,000) (ACCE).

The power of downstream market pull – and well-constructed industrial policies that leverage private capital – is precisely the model that the European Union is now pursuing as it works to establish integrated supply chains that can produce rare earth materials – and rare earth permanent magnets in particular – to supply its rapidly growing electrified vehicle manufacturing base. The twelve E.U. governments that joined forces with private sector companies from around the world to fund the establishment of a resilient European supply chain for batteries ("The Battery Alliance") have already contributed €2.9 billion in funding and are continuing to pursue the deployment of additional investment at an unprecedented level. This government funding is expected to unlock an additional €9 billion in private investments, i.e., more than three times the public support (European Commission – Press release 2021). In my view, Europe is on a path to success in this effort.

5.1.1 History: Early Days of Rare Earths Industrialization in the U.S. and Europe

It is useful to examine how the rare earth industry developed over the past century and, especially, the factors that drove its development. The early industrial history of rare earths is mostly about their pyrophoric properties. The first industrial application dates to the 1880s, when Carl Auer von Welsbach invented gas mantles using 1% cerium. By the 1930s, over a billion gas mantles had been sold globally to illuminate cities around the world. The second application was the blending of rare earths with 30% iron to develop the flintstone (Gupta and Krishnamurthy 2005, 42–43). Later, in WWI, they were used in fuses and explosives (Gilbert et al. 1950).

The period of the densest scientific advancements for rare earths was during World War II's Manhattan Project. The periodic table's lanthanides (a.k.a. rare earths) and actinides (naturally radioactive elements) have very similar chemical properties. Scientists trying to build technologies for the production of actinides needed a safe modeling system. They realized that they could develop their flowsheets using lanthanides as the model system. As a by-product of this effort, separation processes for rare earths were advanced significantly. The first international rare earth patent was filed 1950 by Rare Earths Inc. of New Jersey (Hurst 2010), regarding the processing of monazite sands to extract thorium and rare earth elements.

The mineral resource foundations of the industry generally fall into three categories: (1) bastnaesite and other hard rock fluorocarbonate minerals; (2) monazite found in phosphate hard rocks and as byproducts from the processing of certain heavy mineral sands; and (3) ionic adsorption clays. Certain other phosphate deposits, such as apatites mined for fertilizer production, contain meaningful grades of rare earths but they have yet to be proven as commercially reliable sources of rare earths. It is generally accepted that rare earth production in the U.S. began with mining bastnaesite in the large carbonatite deposits at Mountain Pass, California. In Europe, between 1950 and 1994, the French company Rhône-Poulenc (now Belgium-based Solvay) processed monazite sands to extract rare earths at its La Rochelle plant in France.

Monazite is a rare-earth phosphate mineral that is often a waste ore or a by-product from heavy mineral sands mining aimed at producing zirconium and titanium minerals (Saxon 2020). Monazite is attractive as a rare earth feedstock because its relative distribution of valuable magnetic rare earth elements is typically higher than in bastnaesite and other rare-earth-containing minerals. The challenge to monazite mineral processing? It also contains relatively elevated levels of radioactive uranium and thorium. Those elements can be readily dealt with, of course, but their presence can add high costs and regulatory hurdles to rare earth processing.

5.1.2 The Advent of Rare Earth Mining in the U.S.

In 1952, the Molybdenum Corporation of America acquired the Mountain Pass, California mining claims and began mining rare earths. In 1974, it changed its name to Molycorp. The geology of the deposit's primary mineralization is limited to light rare earths and trace amounts of others. Cerium and Lanthanum make up over 80% of its rare earth distribution. At times, the company did produce small amounts of middle rare earths (Samarium, Europium, and Gadolinium) and minimal quantities of the heavy rare earth Terbium. Beyond that, however, this facility was a light rare earth producer and was never a significant commercial supplier of value-added rare earth materials, other than Europium-based materials for display phosphors. Its primary commercial priority was Cerium compounds for glass polishing and Lanthanum oxide used in petroleum refining catalysts for the production of various gasoline products.

The first rare earth permanent magnets using Samarium and Cobalt were developed in the 1960s from a collaboration between scientists at the Wright-Patterson Air Force Base and the University of Dayton (Ray and Strnat 1972). While these magnets were powerful, they were predominantly used in military applications as they were too expensive for most commercial applications. Mountain Pass produced some of the first Samarium used in these magnets.

The relatively slow rate of growth of rare earth production changed in the 1960s with the advent of color television. It was discovered that the middle rare earth Europium emitted a bright red light when bombarded with electrons, and it began to be integrated into the cathode ray displays of early color T.V.s, ushering in the age of at-home technicolor. The Mountain Pass deposit had a sufficient quantity of Europium to meet this early demand and began separating high value Europium for this nascent market. The small amounts of Gadolinium that Mountain Pass also produced were used in medical scanners and other applications.

Moreover, but parenthetically, Molycorp later operated another rare earth processing facility in Colorado. In 1985, when Yttrium supply was constrained, Molycorp and Sumitomo agreed with Toronto-based Denison Mines to extract the heavy rare earth Yttrium in Louviers, Colorado using tailings from Denison's uranium mine in Ontario, Canada. As the Yttrium supply crunch eased and prices fell with the advent of South China clays, the economics of the facility degraded, and the short-lived rare earth processing there was ultimately shelved.

5.1.3 Rare Earth Industrialization in the People's Republic of China

Not too long after rare earth production began ramping up in the U.S., China began to focus on its, relatively sizeable, rare earth resources. Xu Guangxian, who is considered the father of China's rare earth industry, went to Washington University in St. Louis, Missouri, to conduct graduate work in chemistry in 1946. He finished his Ph.D. at Columbia University, returned to China with the outbreak of the Korean War in 1951, and went on to work on China's nuclear program in 1956. Eventually, Xu went to work on rare earth separation and led the development of the ion exchange separation process for rare earths.

In 1952, the PRC established the General Institute for Non-Ferrous Metals (Hurst 2010). In 1957, rare earth concentrate production began at Bayan Obo in Inner Mongolia. Rare earths were produced as a by-product of iron ore mining at the Bayan Obo mine from the tailing of iron & steel production in Baotou, in Inner Mongolia, so the all-in production costs were quite low. In 1963, China established the Bayan Obo Research Institute in Mongolia.

In the years following the establishment of formal government relations between the U.S. and China in 1972 – particularly during the years of Deng Xiaoping's emergence as Paramount Leader from late 1977 onward – the two countries pursued a scientific transfer program that facilitated academic exchanges of Chinese science students to U.S-based universities and institutions. Signed in 1979 by President Carter and Premier Deng Xiaoping, the *U.S.-PRC Agreement on Cooperation in*

Science and Technology launched technology transfer and commercialization programs between the two nations in the fields of education, energy, agriculture, and space (U.S. Department of State 2006). China had obtained Most Favored Nation status from the US (Tan 1990) and, among other actions, established the Chinese Society of Rare Earths. The research-focused bilateral relationship between the PRC and the U.S. was so strong that the first journal published by the Chinese Society of Rare Earths was printed both in Mandarin and English.

In 1983, a Chinese company filed that country's first international patent application regarding rare earths (Mills 2019). In 1984, the U.S. taxpayer-funded National Center for Industrial Science and Technology Management Development in China's Dalian University was expanded in order to help educate future managers of Chinese state-owned companies (Atkinson 2018). In 1985, The Chinese Ministry of Finance, Customs & Taxation began offering tax rebates to Chinese companies and joint ventures for rare earth exports (Shen et al. 2020). This inevitably drove growth and further capital for rare earth production and exports.

Baotou Research Institute of Metallurgy was renamed the Baotou Research Institute of Rare Earths in recognition of the PRCs focus on rare earths. Today, this facility is the largest research organization in the world dedicated to rare earths. Between 1986 and 1987, two other major academic research facilities were established to focus on rare earths – first the China Rare Earth Information Center and then the Changchun Institute of Applied Chemistry. How dedicated was the PRC to educating an entire generation of technical talent focused on rare earth production and related technologies? According to the Chinese Society of Rare Earths, over 3000 scientists and engineers are working today on rare earths in China. No other nation has come close to developing such an overwhelming intellectual and technical resource focused on rare earths and the technological applications they enable.

5.2 The Making of Rare Earth Supply Chains: 1985–2005

The private sector drove the development of the nascent rare earth industry in the U.S., France, and Japan, with Molycorp, Rhone-Poulenc, and Santoku all using nitric-based solvent extraction separation processes. Processing in a nitric acid environment requires high grade stainless steel which, at the time, was virtually impossible to source (or, at best, extremely expensive) in China. The Chinese rare earth R&D turned to hydrochloric acid processing, which is compatible with inexpensive PVC plastics. To this day, every rare earth plant in China uses hydrochloric acid processing in PVC vessels. Growth was relatively anemic, punctuated by occasional price spikes in individual rare earths such as Yttrium and Samarium in the late 1980s. With little attention paid to the industry by Western governments, China established and started to grow its rare earth industry, capitalizing on its advantages of low-cost by-product raw materials, lower capital and operating expenses of hydrochloric acid separation and ample availability of rare earth processing skills.

5.2.1 Inflection Point: Regulatory Changes in the West

In fact, Western governments took actions that, over the following two decades, greatly facilitated a dramatic decline of rare earth production in the U.S. and Europe. In the early 1980s, the U.S. Nuclear Regulatory Commission (NRC) and the International Atomic Energy Agency (IAEA) tightened transportation standards on low levels of uranium and thorium, classifying them as source materials (United States Nuclear Regulatory Commission 2022). This considerably increased the cost of handling, transporting, and storing, or disposing of these low-level materials.

Naturally, one of the outcomes of this policy was that U.S. and French rare earth separation companies eventually ceased accepting monazite sands because their thorium and uranium concentrations were sufficient to be classified as source material. For example, in the mid- '90s, Rhone-Poulenc stopped processing monazite for rare earths in France because it was no longer able to dispose of the radioactive thorium produced as a result. Around that time, it also ceased rare earth processing from monazite at its Freeport, Texas plant. The French state-owned Rhone-Poulenc was renamed Rhodia, and it soon reached agreement with several entities and with the Baotou Rare Earth Development Zone in Inner Mongolia to build a rare earth production facility there (Schulz 2017), using feedstock mined in China. One year later, Rhodia sold its rare earth metals and alloys assets in the U.S. to Santoku America. Eventually, Rhodia was acquired by Solvay (Solvay 2011).

At Mountain Pass, radioactive issues also significantly impacted Molycorp. The company was then producing a 60% bastnaesite concentrate (on a rare earth oxide basis) and produced a lot of relatively low-purity rare earth products. Increasingly, its cost structure compared unfavorably to the Chinese and Molycorp's business case became difficult to justify. Then, in the late mid-1990s, Molycorp experienced several breaches in its wastewater transport pipeline that sent salty wastewater to evaporation ponds near the Mojave National Preserve. As this pipeline was being cleaned, it over-pressurized and ruptured. There were a few additional pipeline leaks during this period. Some of the mineralized scale that would build up in the pipeline's inner surface contained small amounts of naturally occurring radioactive materials (NORM). As this material was removed, some NORM residue flowed through the pipeline and was part of these spills. California regulatory authorities took action against Molycorp, as a result of the spills. Then, in 1998, faced with the likelihood of additional spills from the troubled pipeline, Molycorp elected to shut down rare earth processing.

5.2.2 China's Rare Earth Industrial Policy

During the first five years of the NRC/IAEA's regulatory phase-in, between 1983 and 1988, PRC-based rare earth facilities witnessed a doubling of demand for their products from Chinese customers, and their production of rare earths quadrupled

(Shen et al. 2020). By the mid-to-late '90s, the global market for rare earths was in the order of 20,000–30,000 tonnes per year (REO). With Mountain Pass offline, China quickly took up the slack, as its rare earth separation and processing capacity was already quite large, and prices continued to come under considerable downward pressure as a result of unrelenting competition by the large number of producers in China. As the midstream of the rare earth value chain became localized in the PRC, specialty component purchasers in Japan, the U.S., and the E.U. of these PRC-produced materials became dependent upon Chinese rare earth compounds. This created the perfect conditions for the rapid creation and expansion of domestic Chinese specialty metallurgical facilities and advanced product manufacturers higher up in the value chain.

This shifting of the global rare earth landscape occurred just as Neodymium-Iron-Boron (NdFeB) permanent rare earth magnets were being widely adopted as critical components in many advanced technologies. By then, the NdFeB magnet not only had become commercial, but production and use had reached critical mass. My company at the time, Advanced Material Resources (AMR), was in fact making neodymium metal and exporting it to Germany and Japan for magnet production by Hitachi and Vacuumschmeltze.

China's focus on developing its rare earth industry was no secret. It was well communicated and perfectly understood among customers and governments in the United States, Japan, and Europe. Western governments watched somewhat passively while China's government actively worked to establish global dominance in rare earth production, the development of technologies enabled by rare earths, and the attraction of downstream value-added production capacity of components and systems that depended on rare earths. The size of the Chinese market, coupled with secure access to rare earths and associated technologies, were very strong driving forces in the migration of supply chains in electronics and automotive to China. When thinking of the difference in degrees between the strategic approaches to rare earths between the PRC and the U.S. – at least concerning industrial policy – Andy Warhol's quote comes to mind: *"They always say time changes things, but you actually have to change them yourself"* (Warhol 1975).

The development of the rare earth industry in China was not all smooth sailing. In fact, it has experienced periods of extreme dislocation. By the early '90s, the vast majority of rare earth producers in China were highly unprofitable. Even small towns had been financing construction of separation plants for over a decade as they were trying to create employment for farmers vying to join the growing middle class. Molycorp was struggling with regulatory and profitability challenges. Rhone Poulenc was facing raw material supply issues and NORM management challenges and decided to switch from monazite to Rare Earth Carbonate from China. As all levels of governments in China focused on attracting foreign direct investment during the '90s, a platform for joint ventures was developed with the establishment of the Baotou Industrial Development Zone. This was specifically designed to attract non-PRC commercial partnerships and investors for this capital-intensive extraction opportunity (Baotou Municipal Government 2014).

In 1999, China's Ministry of Commerce introduced export quotas in an attempt to support export prices. AMR received a comparatively large number of export quotas, sufficient to export the majority of its production from its two plants. AMR's focus on value-added specialty materials had resulted in consistent profitability while its two plants in China were the highest unit value exporters. It was a good time for a new, export-focused player in the industry with China as a base as long as there were sufficient export quotas.

5.2.3 The Rise of the Global-Minded Supply Chain Business: Neo's Corporate Ancestor

In 1992, Peter Gundy, Roco Rossi, and I started working on building a rare earth materials business, and we saw a twofold opportunity in the PRC. First, the byproduct economics of joint rare earth and iron ore mining at Baotou was attractive and economically rational. Second, the discovery and development of ionic adsorption clays in southern China presented a unique opportunity for low-cost extraction of heavy rare earths. John Goode and Kilborn Engineering provided sound technical advice and narrowed down the list of possible partners in China, while Pierre Falconet, a retired Rhone Poulenc marketing executive, made sense of rare earth markets. Carter Trimble, Molycorp's former Director of R&D, provided rare earth-specific technical development and quality guidance.

Peter Gundy, AMR's founder and Chief Executive Officer, was clear about the strategy. We would use China as a low-cost raw material/production base to export high quality refined rare earth materials to compete against the established but struggling producers in France, the U.S., and Japan. We had access to low-cost raw material free of thorium or other environmentally challenging elements. Peter Gundy, a former investment banker and executive with Potash Corp, had in 1993, successfully raised CAD$30 million from global investors and negotiated the acquisition of majority stakes in two Chinese municipally-owned rare earth plants which we started operating in January 1994. No other foreigner had been allowed to negotiate a joint venture in China in this industry before. Les Heymann, AMR's VP of Production moved to Zibo, Shandong in January 1994 for a year, to run the two plants and played a major role in the success of the company.

I started formally working for AMR in 1994 as Business Development Manager, moving to North American Sales Manager, and then to Global Sales Manager, Chairman of the J.V.s, and then VP/General Manager of AMR's Rare Earth Business Unit. In 1997, AMR constructed a plant in Thailand to produce NdFeB isotropic magnetic powders for micro-motor producers in Asia. JVC was our neighbour in the same industrial park and Nidec, the world's largest micromotor producer, a short drive away. In 2000, I became Chief Operating Officer responsible for both AMR's divisions at the time, Rare Earths and Magnetic Materials.

AMR was succeeding with its clear and persistent cost gradient thesis in the PRC and with its customers in Japan, Europe, and North America. As other western industry participants were struggling, we increased our market share, continuously elevating our quality and value proposition. However, by 1997 it was obvious to us that competing against the large number of China commodity rare earth producers was not a wining proposition. We established an R&D laboratory in Abingdon, UK under the leadership of the eminent cerium and zirconium nanotechnologist, Dr. Jim Woodhead. Product development labs were established in the plants in China and AMR's focus shifted to working with and meeting the very demanding expectations of sophisticated customers in Japan, Europe, and the US. The latter half of the '90s continued to see rare earth export volumes from China increasing, while prices dropped. This volume/pricing environment, with the exception of a short-lived spike in the first half of 2001, lasted until 2005 when prices tested the survivability of all the commodity rare earth producers, including those in China. Furthermore, it conditioned western buyers of rare earths on a diet of continuously dropping rare earth prices. So, when the 2011 price spike occurred, it caught numerous companies around the world completely unprepared. It was the first time that the most recent generation of purchasing managers had seen a rare earth price increase.

During the 1990s, AMR was exporting over 90% of its production to Japan, Korea, Europe, and the United States. By the middle of the 2000s, this had dropped to 75%. In 2003, we estimated that 20–25% of AMR's $44-million revenue originated from sales of our rare earth and zirconium-based materials to factories in China, despite this not being a deliberate corporate strategy. By the early 2000s, the PRC's development and rise of the middle-class market became too large to ignore – almost all our international customers started building production facilities in the PRC. Our international customers were establishing plants in China to serve the growing Chinese consumer markets. Supply chains started backfilling as a result. This is what supply chains do; they respond to demand-pull. When global automotive manufacturers, such as Audi and Honda, were building plants in the PRC, and Tier 1 OEMs followed them, so did most of their suppliers. These supply chain forces that pull together geographically consecutive nodes of the supply chain are a principle that has been repeated throughout my whole career. By the end of the 2000s, the balance between what we exported from the PRC versus what was consumed domestically was almost equal. Our customers and their customers had responded to the Circean call of the largest emerging consumer market in the world. Entire supply chains had followed the VWs, Hondas, Benzs, BMWs, LGs, and Samsungs of the world to China.

5.2.4 *The Story of Magnequench*

In the two decades preceding the merger of AMR and Magnequench that created Neo Material Technologies, the pendulum-like ownership story of the intellectual property of the original neodymium-iron-boron magnet has been the subject of extensive analysis by the rare earth/magnet industry as well as in the news media and political circles. Allow me to share my view, given my front-row seat to the action.

In the 1970s, the Samarium-Cobalt (SmCo) magnet was discovered (Ray and Strnat 1972). The instability in 1978 at the Democratic Republic of Congo's predecessor state, Zaire, led to a cobalt price surge of more than 500%, from $5.6 in 1977 to almost $33 per pound in 1979. Simultaneously, there was an issue with the availability and price of Samarium. While the SmCo magnet is of excellent quality and performance, the resourcing challenges of both Cobalt and Samarium made it difficult to guarantee levels of production and stable pricing. This uncertainty troubled original equipment and automotive manufacturers as well as aerospace companies; the supply chain risk was high.

In response, alternatives to SmCo were sought. In 1982, Dr. John Croat, an R&D scientist at General Motors, developed the world's strongest permanent magnet using the rare earth neodymium with iron and boron (NdFeB). At the same time, Masato Sagawa, an R&D scientist at Japan's Sumimoto Special Metals, came up with a strong magnetic material of similar composition. Given the proximity of General Motors' manufacturing to its customers, R&D teams were already fully cognizant of the need for a superior type of permanent magnet for their vehicles. Magnequench was founded as a division of General Motors in 1986, with the sole purpose of producing and commercializing NdFeB magnetic powders and magnets for vehicles. Magnequench filed for patent protection first in the U.S., while Sumitomo filed first in Japan. When Magnequench and Sumitomo came to an agreement to cross-license, G.M. took the rights to isotropic powders for bonded NdFeB magnets, while Sumitomo took the rights to the much stronger anisotropic (a.k.a. sintered) NdFeB magnet field.

In the years leading up to 1996, General Motors began divesting some of its non-core assets, and Magnequench became one of them. As the deal was marketed in M&A circuits, Archibald Cox Jr., a seasoned investment banker (ex-Morgan Stanley and Credit Suisse) who at the time led the boutique private equity and advisory firm Sextant Group, took a particular commercial interest in Magnequench. The deal that General Motors marketed rather broadly received multiple offers, with two bids at the top of the list: one from a large Japanese customer and alloy supplier to Magnequench higher in the value chain, and one from the Sextant Group/San Huan consortium of investors.

San Huan New Materials, a Sumitomo-licensed sintered magnet producer owned by the Chinese Academy of Sciences, approached Sextant. San Huan hoped that financial participation in this buyout deal would create opportunities for closer supply chain integration with Magnequench and would increase the collective competitiveness of its product offering. Archibald Cox committed his own money to this deal and as is customary in the private equity business, sought co-investors to form a consortium. A number of U.S. institutional investors participated, including the Quantum Fund, the Michigan Police and State Employee's Retirement Systems, and Fireman's Fund Insurance among other American institutional investors. Hong Kong-based Onfem Holdings, primarily active in real estate and affiliated with the

PRC's State Nonferrous Metals Industry Administration, also joined the consortium. Given Magnequench's intellectual property leadership in the bonded NdFeB permanent magnet market – a small fraction of which was used in defense applications – the deal required regulatory approval by the U.S. federal government's Committee on Foreign Investment in the U.S. (CFIUS). The deal went through because the majority of shareholders were American, and the Committee's assessment process found no overriding national security risk (U.S. Senate 2005).

Once the deal went through, Archibald Cox was nominated by Magnequench's new shareholders as Chief Executive Officer. The position of Magnequench's board Chair was taken by San Huan New Materials chairman Zhang Hong, the former son-in-law of former Chinese president Deng Xiaoping. Post-retirement, Zhang Hong became the Director of the Research and Development Bureau of the Chinese Academy of Sciences. Together, Archibald Cox and Zhang Hong set the company on track to capture the rising tide of the U.S. tech boom in the late 1990s. This included significant acquisition pursuits, such as the German magnet producer Widia and the French sintered magnet producer Ugimag, increasing the debt leverage of the business significantly to finance these acquisitions. Soros Fund Management and Onfem Holdings exited their investments and were bought out by the company through a leveraged recapitalization of its balance sheet in 1999.

At this point, with Onfem's exit, American shareholders had the majority of ownership. Raising more debt, the company financed the acquisition of Ugimag's sintered magnet factory in Valparaiso, Indiana, and also expanded its global capacity to address new high-growth customers in the PRC by opening a plant there in 2000. By the time the stock market's tech bubble deflated in 2001, many of Magnequench's American customers were suffering financially, and Magnequench's U.S. revenue had decreased significantly. At the same time its two major customers, Seagate and Western Digital, moved their hard disk drive (HDD) production to Asia. In an environment of poor labour relations at its Indiana plants with a highly leveraged balance sheet requiring urgent cost management and adaptation to the new market realities, the plant in Anderson, Indiana, was shut down as its unit costs were prohibitive.

A new magnetic powder plant was built in Tianjin, China, which would make it easier to serve Magnequench's primary markets in Japan, China, and the rest of Asia. While additional cost streamlining efforts were made at the Valparaiso plant over the next two years and a new low-cost magnet facility in Juarez, Mexico was established, the unit economics and the prices of magnets at the time could not justify this business case. Shutting down American operations was a matter of financial survival. This led Magnequench to follow its customers to Asia by moving magnetic powder production to China and sintered magnet production in Singapore (Freeman 2009).

By early 2005, AMR and Magnequench had already been in negotiations for a couple of years regarding possibly deepening their partnerships beyond supply chain arrangements. There was a natural complementary fit between the two companies: AMR was not large enough yet to be global and lacked the critical mass necessary to deal with global supply chains; and Magnequench's punitively leveraged balance sheet required serious repair.

Against this backdrop, AMR and Magnequench agreed to merge to form a new company, Neo Material Technologies Inc., headquartered in Toronto and maintaining AMR's Toronto Stock Exchange listing (TSX: NEM). Notably, the deal did not include Magnequench's sintered magnet assets. It was a difficult merger deal to structure from a financial engineering perspective. Eventually, the Michigan-based hedge fund Stark Investments invested $50 million in the form of a convertible bond, National City (a banker to Magnequench) provided $50 m of senior bank debt, and an additional $70 million was raised in equity to de-leverage the merged company. Archie Cox was appointed Chair of the board, Peter Gundy Vice-Chair, and I was named Chief Executive Officer. A decade and a half later, Magnequench is a key growth and profitability driver for Neo. The merger was a good call for both partners.

5.2.5 2011–2015: First Attempt of End-to-End Vertical Integration

In a new era of improved unit economics, better R&D competitiveness, and a global footprint, Neo established itself as a unique global rare earth specialty value-added materials company. With strong cash flow and a pristine balance sheet, we grew to eleven plants and 1500 employees globally. Fueled by the PRC's successive reductions in export quotas (and the corresponding extraordinary value conferred to the remaining quotas), panic buying, and an unprecedented rise in rare earth prices, Neo's revenue exceeded US$800 million in 2011, with EBITDA of approximately US$300 million. In early 2012, Neo agreed to a CAD$1.3 billion acquisition offer by Molycorp.

The Mountain Pass mine site closure in 2002 led to an inevitable market share dominance by PRC-based producers. China's rare earth export quotas were part of a strategy – at the official level – to prevent the over-exploitation of resources and better protect China's environment. In addition to export quotas, between 2006 and 2008, the PRC government's National Development & Reform Commission (NDRC) imposed production quotas for the separation and processing of rare earths. In 2009, export quota reductions on PRC-produced rare earths sold internationally led to almost a 40% drop in exports. At the time, the PRC's government began communicating its argument that it had to implement quotas to lift prices, capture more tax revenue, and use those revenues to fund environmental remediation, while protecting resources and the environment (Gill 2012). The result was severe supply restrictions and a dramatic price increase by one to two orders of magnitude, depending on the product.

Rare earths have always been considered irreplaceable in the majority of applications. This is substantially but still only partially true. As prices rose to unprecedented levels over a course of months, if not weeks, and, given that our industry works largely without long-term contracts, vulnerable producers in the supply chain without the ability to pass price increases on to their customers faced a serious risk of collapse. At the same time, technologically sophisticated and well-resourced customers embarked on research programs seeking alternative materials and technologies. As a case in point, a collaborative effort launched in 2011 among Neo-Magnequench, Daido Steel, and Honda Motors resulted in the development of very efficient electric traction motors for Honda's electrified vehicles utilizing magnets made by Daido and magnetic powders made by Neo-Magnequench with no heavy rare earths (dysprosium and terbium). Commercial production of these innovative rare earth powders, magnets and motors started in 2017.

Markets, corporate investors, and the news media were convinced that relatively high rare earth price levels were here to stay. With a new ownership structure that included highly credible investors such as Traxys, Goldman Sachs, Resource Capital Funds and Pegasus, Molycorp pursued an IPO with a listing on the New York Stock Exchange in 2010. For Molycorp, the rising prices of rare earths in the global markets and the eventual 2011 spike were beneficial to its business case, resulting in a market capitalization in excess of $7 billion in 2011 and expectations for significantly higher valuations.

The ripple effect of this IPO and Molycorp's rising share price – buoyed by rising rare earth prices – was the corollary establishment of many new junior rare earth exploration companies that managed to secure financing and, relative to what the industry was used to, large M&A activity. Molycorp acquired Santoku's rare earth metals and alloy facility in Tolleson, Arizona in 2011, and then the rare earth processing facility in Sillamäe, Estonia, in April 2012. In June 2012, Molycorp acquired Neo.

The PRC's export quotas reduction triggered concerns by many U.S. stakeholders about the security of supply for rare earths. This only worsened with the 2011 rare earth price spike. Molycorp's acquisition of Neo closed in the second quarter of 2012. Molycorp raised $650 M in a senior secured bond carrying a 10% interest to finance the deal. This eventually became a serious cash drain. The industrial logic behind the merger was solid: putting together an excellent rare earth resource upstream with a leading mid-level rare earth processor to achieve Molycorp's senior management team's vision of vertical integration, or "mine-to-magnet," made sense. The debt load, and corresponding debt servicing drain on cash flow, and its inability to bring the ambitious flow sheet into production on time and budget, became insurmountable obstacles for Molycorp.

On the export quota front, the PRC government argued that these policies to reduce export volumes were necessary for environmental and resource management, particularly in the upstream and midstream. This reasoning was not accepted by the US, the EU and Japan. The culmination of this dispute was a U.S.-led trade complaint to the World Trade Organization, supported by the E.U. and Japan, arguing against the PRC's policy of export restrictions. There was a partial settlement of the dispute through the WTO process, and the PRC indeed withdrew export quota restrictions between January and May of 2015. Predictably, this resulted in a substantial collapse in prices, forcing Molycorp to seek bankruptcy protection in 2015.

Of note, here, is the fact that during the entire period that Neo was owned by Molycorp, the business was handsomely profitable and cash flow positive; just not sufficiently to finance the unsustainable losses in the rest of Molycorp's business.

During this time, there was an interesting parallel in China's fiscal policy development. While Molycorp was going bankrupt, the PRC was developing a more calculated policy response. In 2015, the PRC introduced a new tax on rare earth resources – 27% for medium and heavy rare earths and 11.5% for light – while it reintroduced a little-used strategic stockpile for rare earths, stepping in the markets buying rare earths when prices were "too low" and selling rare earths when prices were "too high". The strategy has been effective on multiple levels: first, it ensured that local governments would have a revenue source for setting and maintaining a high standard of environmental measures; second, it kept prices from falling below the breakeven point of PRC-based companies. Third, as a result of the "substitution" lessons learned in 2011, China prevented prices from rising to the point where "*smart engineers and scientists with deep pockets look for alternatives; because they will, eventually, find them*", as I commented at the annual Roskill-Metal Events rare earth conference in Hong Kong in 2017.

The PRC's policy response was not the result of a sophisticated geopolitical game-theory simulation, but rather the very plain and simple appreciation for methodical industrial policy interventions to ensure that local and global manufacturers in China would have the necessary domestic supply chains needed to produce cars and electronic devices to meet the demand of China's rapidly growing middle class, while providing an ever-growing supply of middle-class jobs. China has a recognised advantage in rare earths and associated technologies and uses every tool in its well constructed toolbox to protect this advantage.

Following its filing for bankruptcy protection in 2015, a year-long court process led to Molycorp's emergence from bankruptcy with Oaktree Capital, a creditor whose loans were secured by Molycorp's downstream assets (the assets that Neo brought into the company plus the Silmet rare earth, tantalum and niobium production plant in Estonia) took ownership of these assets in 2016, naming this iteration of the company as Neo Performance Materials. I was asked to serve as Board Chairman, and Neo went on to be listed on the TSX again in 2017. Approximately 30% of Neo was sold to the public in the 2017 IPO. The Mountain Pass mine was put on care and maintenance under receivership until it was eventually auctioned for $20.5 million in the summer of 2017. The new and current owners were an American-led consortium led by M.P. Mine Operations LLC ("MP Materials") with a minority stake held by the Chinese rare earths upstream company Leshan Shenghe Rare Earth Co. Today, MP's Mountain Pass operations mine and produce a rare earth concentrate sent to China for separation. MP Materials' stated objective is to establish rare earth separation, metal/alloy, and magnet production in the U.S.

5.3 2016 To Present: At New Inflection Point of Neo's Growth

5.3.1 Performance Is in the Neo Name

In the 1990s, it became very clear that there is no way AMR could survive by selling commodities – it had to make strategic investments in products higher up the value chain through R&D and working closely with customers. This would be the company's differentiator, and today it has largely succeeded with a business model that captures a value-added margin on top of the underlying commodity price, regardless of whether prices are moving up or down.

The most important privilege of wisdom and perspective accumulated from three decades in the rare earth trenches is knowing where along the value chain one is best positioned to create value. Our team at Neo had worked throughout the complete spectrum: from minerals to advanced materials. With the new corporate entity that emerged out of Molycorp's bankruptcy process, and with the capital support of Oaktree, the company has been able to preserve its most strategic asset: its people. Along the way, much like permanent magnets, some of this global industry's most knowledgeable, competent, and passionate professionals from different career paths have stuck with one another. Whether it was the AMR and Magnequench merger, the growth of Neo Materials, the acquisition of Recapture Metals, the merger with Molycorp, and now the team of Neo Performance Materials and its three divisions – Chemicals & Oxides (rare earths and rare earth specialties), Magnequench and Rare Metals, the core group of lifelong loyal team members keeps growing stronger and larger.

Unlike the previous names of our corporate entity ancestors, "Performance" is in the company name today for a reason: it serves as a key differentiator. Our team at Neo appreciates that supply chains will continue to reconfigure over the next decade with demand-shaping location choices for OEMs, pulling TIER-1s and the rest of their supply chains along. Mines will continue to get developed where it makes economic sense. The rest of the supply chain participants will need to ensure they bring an optimum combination of values (location, capacity, price, performance, quality, ESG practices, reliability, product development, security of supply…to name just a few of these values) to their customers. The few midstream businesses will do their best to adapt their logistics and costs to serve in those conditions while pursuing acceptable margins and return on capital. Most downstream businesses, however, will follow customers wherever they locate at a relatively lower transition cost. And, if they have proprietary technologies and can create unique high-quality/high performance specialty products on which the performance and profitability of their customers depends, they can capture higher margins and growth rates.

5.3.2 *Value in Magnet-to-Mine: Second Attempt of End-to-End Vertical Integration*

Driven by the vocal OEM demand for supply chain optionality, Neo's European and North American growth plan created a new necessity based on an old theme: mine-to-magnet. For Neo to be in a position to minimize the price volatility of permanent magnets for its automotive customers in the long run, it needed to be in a position to reduce the cost volatility of its upstream feedstock. There were five key variables in the company's decision-making equation when pursuing investments in the upstream: first, the price had to be fair; second, the deposit needed to be almost free of Uranium; third, the host country of the deposit had to consider the development of the deposit a matter of national priority; fourth, the infrastructure gap to make the transportation logistics work should be small; and fifth, visibility on the operating economics in that jurisdiction was essential. After a global scan of the available deals at the spring of 2022, the Sarfartoq deposit in Greenland checked all these five boxes perfectly. The deal price was fair, considering that the seller – Hudson Resources – wanted to exit this rare earth asset to focus on its core competency where it already had become successful: anorthosite mining in Greenland. Based on Hudson Resources' geological studies completed a decade earlier, the mineralization of the deposit was rich in magnetic rare earths and exhibited only a tenth of the Greenlandic government's regulatory threshold for radiation in mining. The Government of Greenland had indicated both in public literature and at conferences that it considered this deposit to have potential and that it viewed mining – backed by a meaningful balance sheet and a proven commitment to sustainability – as the core driver of its national economic development strategy (Government of Greenland 2020). The Sarfartoq deposit is on the south side of the Kangerlussuaq Fjord with deep water access that can operate year-round – it will allow for loading beneficiated rare earth concentrate to ships that could offload directly at the port next to the Silmet plant in Estonia. And finally, across the water body, on the north side of the fjord, Hudson Resources had successfully constructed and is operating its White Mountain anorthosite mine, which gives Neo visibility on both CAPEX and OPEX comparable economics.

Neo's renewed conviction for an integrated magnet-to-mine supply chain comes with a new priority objective: relentless pursuit of implementing sustainability in the root design of its new upstream and downstream assets. To the degree possible, the selected assets either have access to installed or planned renewable energy sources. Additionally, as members of the Responsible Minerals Initiative, Neo is developing both the mining project in Greenland and the magnet plant in Estonia with sustainability considerations from the design phase. All this comes as a natural advancement step after Neo's team already institutionalized Neo's global sustainability function, began collecting extensive ESG data, published its inaugural Sustainability Report, and chose Export Development Canada as its financing partner for strategic projects which conditionalizes its lending on ESG performance at the highest global standards. At Neo we believe in the energy-saving effect of rare earth permanent magnets equally as much as we believe in the continuous sustainability improvement of our own operations.

5.4 Public Discourse and Cognitive Biases

Paraphrasing Charles Freeman's quote about Magnequench in his January 2009 article for the Washington Quarterly, the coverage of our industry is *"rich in symbolism, containing as much fable as fact. Globalization and fundamental economics drove manufacturing plants to China*", not a geopolitical weaponization (Freeman 2009). The prevailing media narrative of the industry has been dominated by political passions of geopolitical antagonism between the U.S., the West broadly, and China. This overlooks the fundamental facts on the ground where China has been pursuing a position of leadership in an industry it identified as critical for the development of global industrial manufacturing capabilities and jobs for its growing middle class. China has invested in the development of scientific centres of excellence on rare earth processing and applications, as well as the education of thousands of scientists and engineers who become immediately productive when they are hired by rare earth producers. It continues to provide generous financial incentives and credits to companies securing resources beyond its borders and pursuing novel R&D and capacity expansion within its borders.

Popular art is usually a reflection of contemporary social or political struggles. In the case of the public opinion on rare earths in the United States, the rare earths industry has already been privileged to have its Hollywood moment. In the Netflix series "*House of Cards,*" the sixth episode of the second season involves the United States going through an energy crisis due to a Chinese ban on Samarium 149 – a rare earth isotope used as absorbing material in nuclear fuel control rods. As far as the printed word goes, "*All the Devils are Here,*" the latest murder-mystery novel taking place in Montreal and Paris by Louise Penny, the popular Canadian novelist, is about – what else? – rare earths. A bit far fetched but rare earths have safely entered the popular fiction domain. Most recently, the 2021 Hollywood movie "*Don't Look Up*" made reference to the – fictional – opportunity of space mining a meteorite for its trillion dollars in value of rare earth minerals it contained. For the longest time, I would joke that a video game production on rare earths was truly the only thing missing. To no surprise, my younger friends pointed out to me that the e-sports industry was already ahead of me, producing "Call of Duty: Black Ops 2" in 2012 with a rare earth minerals quest background theme.

In this environment, very frequently, rare earth supply security is inextricably linked to national defense. While there are key applications for rare earths in defence systems, the total annual demand by U.S. defense technologies is hardly sufficient to baseload a single world-scale rare earth production plant. In the end, while any perceived national security risk is justifiably an issue of concern that needs to be addressed, the narrative has to reflect the size of the problem so that the policy response is proportionate and respectful of the taxpayer's money. Policy and industrial strategy need to be informed by a proper understanding of the rare earth market, supply/demand dynamics and supply chain formation.

5.5 Looking to the Future: Capital Formation and Public Policy

5.5.1 Unlocking Value in Upstream & Midstream of Rare Earths

As the saying goes, history does not repeat itself, but it does rhyme. In the case of Neo's latest supply chain initiative to capture the benefits of by-product economics through monazite sands in a partnership with Colorado-based Energy Fuels, history certainly rhymes with how Advanced Material Resources (Neo's corporate ancestor) was initially founded. This time, the by-product economic benefits are not achieved by the production cost gradient between the US/EU and China. Instead, these benefits are achieved using Neo's strategic separation facility in Estonia and the aggressive pursuit of improved unit costs by two U.S. resource transformation and specialty chemicals companies, Chemours and Energy Fuels.

Because monazite is typically a by-product from other mining operations, such as Chemours' heavy mineral sands processing for the recovery of ilmenite and zircon in Georgia, processing monazite sands for the recovery of rare earths, along with the contained uranium, can be done at a relatively low cost since no new mining needs to be licensed, financed, and developed. Furthermore, Energy Fuels is utilizing its existing capacity at its White Mesa Mill to produce the RE Carbonate that our Silmet rare earths facility in Estonia can process, and Silmet will utilize its existing separations facility to process that material. This avoids the need to license, finance, and develop new facilities with these capabilities. As a result, producing rare earths and uranium from existing mining, processing, and separation facilities is not only expected to operate at a relatively low cost, but it also will result in a much smaller environmental footprint as compared to other greenfield projects.

This is an example of how smart development can make up for cost differentials between the West and China, irrespective of geopolitics and by following customer demand: by-product economics, capital-efficient projects, specialty products and highly productive, motivated people. This gives you a fighting chance to compete and some protection from the low-price phase of the cycle. From a competitive perspective, junior mining projects in the West are hobbled by a structural disadvantage in that they tend to apply a mining mindset to an industry that requires more specialty chemical knowledge. It's very difficult to bridge the mining and specialty chemical gap in this industry.

Thus, in my view, producing value-added rare earth materials from monazite resources is a massive step forward for our industry and for consumers of rare earths. Monazite is an excellent and rich source of rare earths, particularly the magnetic rare earths that are highly sought for new electrification applications. Yet, monazites derived from heavy mineral sands have historically not been favored due to the naturally occurring radioactive materials (NORM) they contain. Energy Fuels provides the missing link in solving this challenge. They extract valuable uranium from monazite and put it to good use while also recovering monazite's rare earth content. And, given the relatively significant monazite resources around the world, we expect this collaboration will help Neo expand its production of separated rare earth products in Europe and improve the resiliency of European manufacturing by moving further downstream. That is particularly timely given the increasing demand for rare earth materials across the continent and given the E.U.'s vehicle electrification initiatives, which will require significantly greater supplies of rare earths. We also expect that, when the demand develops in North America, this business model could be put to good use there too, as the policy tailwinds of the Inflation Reduction Act come into effect.

5.5.2 *Public Policy for Energy-Saving Rare-Earth Permanent Magnets*

As the concluding section of this chapter, I would like to leave the reader with a forward-looking thought about the catalytic consequence that public policy can have in capital formation, using the case study of the European Union. Specifically, I will elaborate on the E.U.'s current thinking around public policies designed to help capture the benefits of midstream and downstream co-location in order to establish an E.U. domestic supply chain of rare earths and permanent magnets.

Today, Europe imports the vast majority of the rare earth permanent magnets it consumes. Why? Because China identified rare earths as a strategic industry in the late 80s, included them in its industrial strategy and successive five-year plans over the past four decades, and offered strong financial, technical development, and other government support at every node of the value chain. This included assistance in mining, rare earth separation, R&D on mining/processing/magnet production and applications, municipal and federal permitting, low cost – even free – land, labor cost subsidies, and skills development and training.

As a result, the production of rare earths and magnets in China enjoy significantly lower capital and operating costs, especially by State-Owned Enterprises (SOEs) and, among those, particularly for the six SOEs appointed by the government several years ago to "consolidate" the industry in China. The cost of capital for SOEs is much lower than for other companies anywhere else in the world. In fact, it is practically negative. This allows Chinese producers to operate on razor-thin margins and to dominate global commodity markets through pricing alone. This, naturally, has led to challenges and market failures beyond China's borders.

Exporters of rare earths cannot recover the VAT on exports. Exporters of magnets, on the other hand, can. Thus, magnet producers outside of China that rely on rare earths from China must operate at a 13% cost disadvantage compared to their China competitors. This disadvantage also extends to magnet producers outside of China who buy rare earths from non-Chinese sources, given that China, due to its dominant demand supply positions, effectively sets global rare earth prices.

Because of Neo's strategic rare earth asset in Estonia, it is uniquely positioned to play its part in the re-emergence of a European supply chain that – by extension – would help create domestic jobs in electric vehicle manufacturing. The E.U. has delivered consequential policy work with the European Institute of Innovation and Technology (EIT) and the European Battery Alliance in establishing a globally competitive E.V. battery supply chain. The E.V. drivetrain challenge, and the rare earth supply chains that enable these drivetrains, can be solved as well. Tackling this challenge will require bold, long-range industrial policies. Europe has a significant head start in the effort.

To remain competitive in global rare earth markets, Neo has carved profitable market niches that do not fit the Chinese SOE model. The SOEs are focused on large volume commodity markets and applications. They have not been interested in more value-added specialty materials opportunities and niche markets where innovation and close customer relations are indispensable factors for success. Neo produces customized, value-added rare earth products such as specialty chemicals used in environmental catalysts and electronic devices, and high-performance magnetic powders and magnets for automotive, electronics, home appliances, and any motor application where energy efficiency is a primary attribute. As a result, Neo does not generally compete with the large Chinese commodity rare earth producers.

Constructing a resilient European rare earth and magnets supply chain will take bold industrial policy intervention – just as China did several decades ago with its domestic industry. In my view, there are four moving policy parts that the European Union needs to align. First and foremost, European vehicle manufacturers would have to be incentivized to utilize European rare earth magnets as much as possible in their operations. These companies have been relying on Asian supply chains for their components for a very long time, while they have also been making a great deal of their profits in the Chinese market. European magnets will cost more than those made in China, at least until critical mass is reached and raw materials costs equalise. OEMs and Tier-1 producers will have to accept a modest price premium for their motors and drive trains, reflecting a premium on EU-produced separated rare earths and magnets. If the OEMs commit to this, the rest of the supply chain will follow. Second, companies should be encouraged to collaborate and participate across the value chain – alternative upstream sources, midstream separation, and downstream research and development clusters.

Third, corporate pursuit for circular economics and recycling should be rewarded, as the economics are challenging to work in times of low material prices. Neo has been continuously pursuing recycling technologies and opportunities in the different parts of the supply chain. For example, its Rare Metals Division currently operates a plant in Sagard, Germany that recycles rare metals such as tantalum, niobium, rhenium, and hafnium from aerospace alloy scrap and end-of-life parts, with the final products sold back into the aerospace supply chain. Neo's plant in Ontario, Canada (also part of the rare Metals Division), recycles Gallium and Indium from semiconductor and solar panel scrap. About 20% of the magnet alloys Neo produces and turns into magnetic powders for Daido Electronics in China and Thailand are made from recycled rare earth magnet scrap that Daido returns to Neo. Also, Neo's rare earth magnetic powders that go into magnets for certain electronic devices contain recycled rare earths. Collaboration and incentives will drive efficiencies, reduce GHGs and other environmental impacts, and help the company approach 100% closed-loop recycling. Fourth and final, for private capital to be deployed, a proper return on investment is required. The fundamental way for the E.U. to address this market failure is through a combination of grants supplemented by low-interest credit for CAPEX needs.

In 2022, our industry witnessed an inflection point in EU public policy. The EU Commission's President Ursula von der Leyen committed to introduce legislation – the Raw Materials Act – and create the EU Sovereignty Fund (Breton 2022) for the support of rare earth supply chains in Europe, among other strategic materials value chains. For the public policy leaders in Brussels to reach this decision, they had to look no further than the bold examples of energy saving that permanent magnets have. In the case of electric vehicles, two kilograms of rare earth permanent magnets in the drivetrain motor save 20% of the size of the battery, which is the most expensive component – savings that translate to more than an order of magnitude in weight of lithium, cobalt, and nickel required. In the case of wind energy generation, at least 70% of offshore turbines require an average of eight hundred kilograms of rare earth permanent magnets each. In the case of residential water circulation pumps, the use of our magnets by one of Neo's longest customers prevented 20% in energy losses of this application in Europe – the equivalent of the installed capacity of two nuclear power plants. While it took a lot of effort by the rare earth industry and its customers to explain to EU public policy leaders how critical these supply chains are for Europe to meet its carbon reduction targets, there are happily improvements in the understanding and coverage of the rare earth industry, relative to a decade ago.

5.5.3 Final Thoughts

Whether a cause or a consequence of supply chain forces that attract consecutive nodes of a supply chain, industrial cluster formations have been a recurring theme of my 30-plus years in the rare earth industry. Regardless of the jurisdiction – E.U., the United States, the People's Republic of China, India, Brazil, or Australia – tackling such an industrial challenge requires bold and long-range public policies to succeed. These policies should be fashioned to help the industry reach critical mass and then allow business-academic clusters to lead the innovation required to develop high-margin downstream technologies and products. This is only possible by appreciating and taking advantage of the supply chain forces necessitating the geographic proximity of the industry's midstream, downstream, and eventual customers.

The world will continue to search for smart alternatives, such as unique mineralization that leads to lower cost of production. Product development will lead to higher performing magnets for more energy-efficient motors. Finally, ESG will continue to play an ever-increasing role in these supply chains with sustainability, recycling and circular economy considerations becoming more central and determinant factors in our supply chains. This is why I am bullish and excited about the ever-increasing strategic role that rare earth elements will play in the years ahead as the world decarbonizes and begins to tackle the challenges presented by a changing climate.

References

Atkinson, Robert D. 2018. Who lost China? *National Review* (blog), 26 July. https://www.nationalreview.com/magazine/2018/08/13/us-china-relations-who-lost-them/. Accessed 13 Nov 2022.

Baotou Municipal Government. 2014. Baotou rare earth high and new technology industrial development zone. *Baotou Municipal Government*, 12 February. https://www.baotou.gov.cn/info/1685/16148.htm. Accessed 1 Sept 2022.

Breton, T. 2022. https://ec.europa.eu/commission/presscorner/detail/en/STATEMENT_22_5543

European Commission – Press release. 2021. State aid: Commission approves €2.9 billion public support by twelve Member States for a second pan-European research and innovation project along the entire battery value chain. *European Commission*, 26 January. https://ec.europa.eu/commission/presscorner/detail/en/ip_21_226. Accessed 13 Nov 2022.

Fink, L. 2021. It was Morgan Stanley's financial decision makers ESG virtual conference, 25 February 2021: https://www.cvent.com/events/financial-decision-makers-esg-virtual-conference/archived-70deca1c2a7747d5bba47ec7fff51ad7.aspx

Freeman, C.W. 2009. Remember the Magnequench: An object lesson in globalization. *The Washington Quarterly*, 32:1 (January):61–76. From https://csis-website-prod.s3.amazonaws.com/s3fs-public/legacy_files/files/publication/twq09januaryfreeman.pdf. Accessed 28 Nov 2022.

Gilbert, Paul T., P. Edwin, and Arthur. 1950. Glass electrode. *United States Patent Office* 2668143, filed January 19, 1950, and issued February 2, 1954.

Gill, J. 2012. Chinese White Paper on rare earths signals shift in global market. *China Briefing*, 25 June. https://www.china-briefing.com/news/chinese-white-paper-on-rare-earths-signals-shifts-in-global-market/. Accessed 13 Nov 2022.

Government of Greenland. 2020. Greenland's mineral strategy 2020–2024. *The ministry of mineral resources*, February. https://govmin.gl/wp-content/uploads/2020/03/Greenlands_Mineral_Strategy_2020-2024.pdf. Accessed 13 Nov 2022.

Gupta, C.K., and N. Krishnamurthy. 2005. *Extractive metallurgy of rare earths*. Boca Raton: CRC Press.

Hurst, C. 2010. China's rare earth elements industry: What can the West Learn? *Institute for the Analysis of Global Security (IAGS)*, 1 March. http://www.iags.org/rareearth0310hurst.pdf. Accessed 15 Nov 2022.

Interesse, G. 2022. China considers extending its EV subsidies to 2023 (updated). *China Briefing*, 29 September. https://www.china-briefing.com/news/china-considers-extending-its-ev-subsidies-to-2023/. Accessed 13 Nov 2022.

Mills, R. 2019. How the U.S. lost the plot on rare earths. *Mining.com*, 25 January. www.*mining.com* (blog). Accessed 13 Nov 2022.

Ray, A.E., and K.J. Strnat. 1972. Research and development of rare earth-transition metal alloys as permanent-magnet materials. *Technical Report* AFML-TR-72-202. (August). https://apps.dtic.mil/sti/pdfs/AD0750746.pdf. Accessed 10 Nov 2022.

Saxon, M. 2020. Medallion completes siting study for North American. *Globe Newswire*, 14 October. https://www.globenewswire.com/news-release/2020/10/14/2108195/0/en/Medallion-Completes-Siting-Study-for-North-American-Rare-Earth-Element-Production-Facility.html. Accessed 13 Nov 2022.

Schulz, K. 2017. *Critical mineral resources of the United States. Rare-earth elements*. Reston: U.S. Geological Survey.

Shen, Yuzhou, Ruthann Moomy, and Roderick G. Eggert. 2020. China's public policies toward rare earths, 1975–2018. *Mineral Economics* 33 (1 (1 July)): 127–151.

Solvay. 2011. Solvay to create with Rhodia a major player in chemicals – A global leader in their businesses, *Globenewswire*, 4 April. https://www.globenewswire.com/en/news-release/2011/04/04/1692625/0/en/Solvay-Press-release-Solvay-to-create-with-Rhodia-a-major-player-in-chemicals-a-global-leader-in-their-businesses.html. Accessed 13 Nov 2022.

Tan, Q. 1990. The politics of U.S. most-favored-nation treatment to China: The cases of 1979 and 1990. *Journal of Northeast Asian Studies* 9 (March): 41–59. https://doi.org/10.1007/BF03025175.

U.S Senate. 2005. Hearing on foreign investment in the US. *Committee on Banking, Housing, and Urban Affairs*, 20 October. https://www.banking.senate.gov/imo/media/doc/kimmitt.pdf. Accessed 2 Nov 2022.

U.S. Department of State. 2006. *U.S. China Science and Technology Cooperation (S&T Agreement): Report to Congress*. 30 December. https://2001-2009.state.gov/g/oes/rls/or/2006/96328.htm. Accessed 1 Oct 2022.

United States Nuclear Regulatory Commission. Last reviewed 2022. 40.22 Small quantities of source material. *U.S.NRC*, 9 May. https://www.nrc.gov/reading-rm/doc-collections/cfr/part040/full-text.html#part040-0022. Accessed 1 Nov 2022.

Warhol, Andy. 1975. *The philosophy of Andy Warhol*. Orlando: Harcourt.

Constantine Karayannopoulos serves as Neo Performance Materials' (TSX: NEO.TO) President and Chief Executive Officer since July 2020. He is also a Director of Neo and serves on the Health, Environment, Safety and Sustainability (HESS) Committee. Mr. Karayannopoulos has worked in senior executive positions at Neo and its predecessor companies for more than 30 years, including as Director, President, and Chief Executive Officer of Neo Material Technologies ("NEM") (TSX:NEM) from 2005 until NEM was acquired in June 2012 for US $1.3 billion. Over the past three decades, as one of the longest serving executives in the rare earth, rare metal, and battery materials industries, he has worked with multiple governments in Europe, North America and Asia to converge public policy and industrial economics. Karayannopoulos was also co-founder and non-executive Chairman of the Board of Neo Lithium Corp. (TSX:NLC), which owned the second highest grade lithium brine deposit in the world, in Argentina. As non-executive Chairman, until its sale for $900 million in 2021, he oversaw the governance of both the mining project and the processing plant being built. He is a member of the Advisory Board at the University of Toronto's Department of Chemical Engineering and Applied Chemistry, a Director of the Canada China Business Council, and a member of the Advisory Board of a private merchant bank that specializes in energy and technology transactions. He holds Bachelor and Master of Applied Science degrees in Chemical Engineering from the University of Toronto.

Vasileios Tsianos supported Constantine Karayannopoulos' primary authorship of this book chapter. He serves as Director of Corporate Development at Neo Performance Materials, leading new business projects, the investment in the Greenland mining development and government relations. He also serves as Assistant Corporate Secretary, supporting the corporate governance function of Neo's Board of Directors. Concurrently, Vasileios serves as Board Director and Executive Committee Member of the European Union Chamber of Commerce in Canada, and as Board Director of the Global Rare Earth Industry Association (REIA) and Chair of REIA's Public Policy Committee, and as President of the Hellenic Canadian Board of Trade. Before joining the rare earths industry, he served as Management Consultant at StrategyCorp, where he advised clients in highly-regulated industries on corporate development strategy. Prior to that, he served as Venture Partner at NextGen Venture Capital, as Vice President of Corporate Development at Keela.co, and as Investment Banking Analyst at IJW & Co. He has also served as campaign chair and senior advisor to several political candidates and elected officials in Canada, the European Union, and the United Nations. In 2015, he served as Head of Delegation for a major EU interest group at the COP21 UN Climate Summit in Paris. In the past, he has been recognized as one of the 100 Future Global Leaders by the Web Summit, a Greek America's 40 Under 40, a Fellow of the Royal Society of Arts, and a Global Shaper of the World Economic Forum. He holds a Bachelors in Chemical Engineering from the University of Toronto and a Masters in Finance from IE Business School.

Chapter 6
The Paradox of Green Growth: Challenges and Opportunities in Decarbonizing the Lithium-Ion Supply Chain

Chris Berry

Abstract Global decarbonization is a megatrend destined to impact multiple sectors of the economy. The commodity sector in particular looks set to benefit, given the acute need for raw materials to feed the energy transition. This global transition will take decades and while an "all of the above" energy policy will be necessary, the lithium-ion battery promises to play a central role in the decarbonization of the transportation and grid scale energy storage sectors. With lithium-ion battery costs falling steadily since the early 1990s, many believe that the technology promises to become even more ubiquitous, helping to electrify the global economy. Recent exogenous shocks such as the US-China trade war, COVID-19, and the Russian invasion of Ukraine have underlined the fragility of global energy supply chains and the need for a resilient global energy infrastructure. While solutions to the exogenous shocks seem straightforward, there is an increasingly evident paradox. To achieve significant decarbonization and electrification goals, more, not fewer, battery raw materials will be required and the energy used to produce these materials will almost certainly originate from fossil fuels, possibly slowing the decline of the carbon intensity of industry. Innovation along the supply chain is the only realistic way to achieve large-scale decarbonization. This will require R&D, necessitating a coordinated response between the public and private sector as supply chains evolve. This chapter looks at the geopolitical, economic, technological, and legislative challenges and opportunities that green growth presents in the lithium-ion battery ecosystem as the industry embarks on a rapid growth phase.

Keywords Lithium · Cobalt · Nickel · Lithium-ion battery · Supply chain · Legislation

---

C. Berry (✉)
House Mountain Partners, LLC, Washington, DC, USA
e-mail: cberry@house-mountain.com

S. Kalantzakos (ed.), *Critical Minerals, the Climate Crisis and the Tech Imperium*, Archimedes 65, https://doi.org/10.1007/978-3-031-25577-9_6

6.1 Introduction

Global decarbonization is a megatrend destined to impact multiple sectors of the economy. The commodity sector in particular looks set to benefit, given the acute need for raw materials to feed the energy transition. This global transition will take decades and while an "all of the above" energy policy will be necessary, the lithium-ion battery promises to play a central role in the decarbonization of the transportation and grid scale energy storage sectors. With lithium-ion battery costs falling steadily since the early 1990s, many believe that the technology promises to become even more ubiquitous, helping to electrify the global economy. Recent exogenous shocks such as the US-China trade war, COVID-19, and the Russian invasion of Ukraine have underlined the fragility of global energy supply chains and the need for a resilient global energy infrastructure.

While solutions to the exogenous shocks seem straightforward, there is an increasingly evident paradox. To achieve significant decarbonization and electrification goals, more, not fewer, battery raw materials will be required, and the energy used to produce these materials will almost certainly originate from fossil fuels, possibly slowing the decline of the carbon intensity of industry. Innovation along the supply chain is the only realistic way to achieve large-scale decarbonization. This will require R&D, necessitating a coordinated response between the public and private sector as supply chains evolve. The chapter examines the geopolitical, economic, technological, and legislative challenges and opportunities that the green transition presents in the lithium-ion battery ecosystem as the industry embarks on a rapid growth phase.

Secure access to a supply of battery metals is vital to electrify global transportation via the lithium-ion battery. The boom in the battery metals since late 2020 has provided all stakeholders along the lithium-ion battery supply chain with a front row seat to extreme pricing volatility. This pricing volatility is a nightmare for corporations planning major capital investments and can scare away traditional investors who are an important source of capital (Figs. 6.1 and 6.2).

The roots of this volatility stem from a host of factors including US-China trade war friction, COVID-19 supply chain fragility, and fallout from the Russian aggression in Ukraine. Demand for significant amounts of raw materials to fuel the energy transition is clear, but the lack of pricing transparency and misallocation of capital in past booms such as 2016–2018 put relentless downward pricing pressure on lithium and cobalt, two metals vital to successful decarbonization. This temporarily slowed electric vehicle (EV) supply chain development, despite EV sales continuing their steady upward march across the globe. The market has recovered despite the shock from COVID-19 and in 2021 EV sales doubled to over 6.6 million with more sold each week than in the whole of 2012 according to the IEA (International Energy Agency 2022). Consensus growth forecasts are as high as 30% per year in EV sales to 2030 (Walton et al. 2020). This has been encouraged by governments across the globe promoting EV adoption via a carrot and stick approach. The United States, for instance, offers consumers subsidies to purchase EVs (United States Congress 2022a) while the European Union punishes automotive manufacturers if

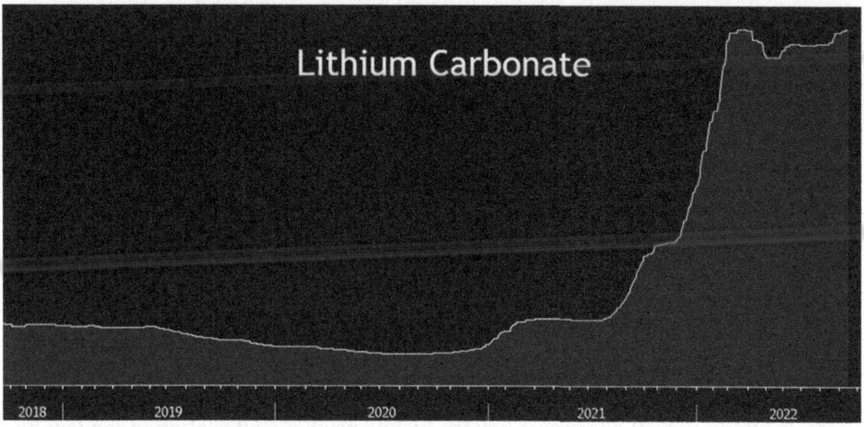

Fig. 6.1 Lithium carbonate price October 2018–October 2022. (Source: Bloomberg, Asian Metal)

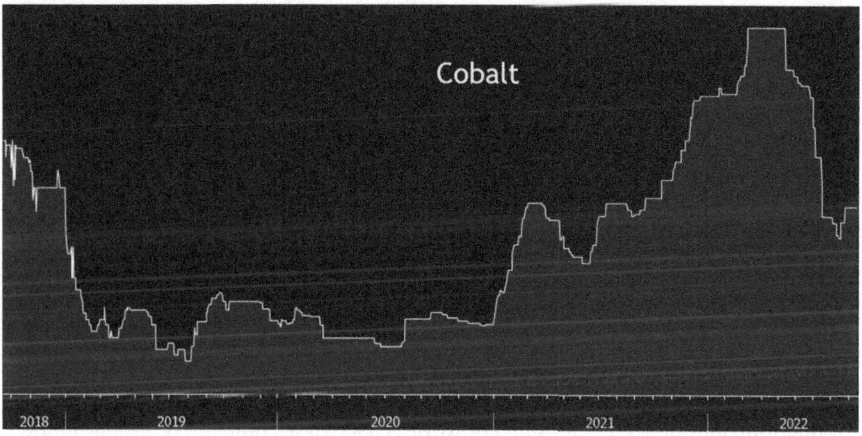

Fig. 6.2 Cobalt price October 2018–October 2022. (Source: Bloomberg, CME)

their carbon dioxide emissions per kilometer driven exceed a certain threshold (in addition to its own consumer subsidy programs) (European Commission 2020).

Decarbonizing transportation is vital in the fight against climate change; however, it raises a significant paradox for all stakeholders along the battery metals supply chain. Successful decarbonization on an economy-wide scale will require *more,* not fewer, raw material supplies and in addition to this, the development of larger lithium-ion supply chains will necessarily be powered by fossil fuels in the near-term. This has direct implications for carbon dioxide emissions in the transition to a greener economy.

As an example, lithium hydroxide produced from hard rock mining of spodumene ore is up to seven times more carbon dioxide-intensive relative to other processes due to the need for energy, roasting, and chemical conversion (Grant et al. 2020). More EVs on the world's roads requires more batteries which requires more raw material. Simply mining less lithium or cobalt is not an option as this would see the automotive industry fall well short of EV sales goals. Fortunately, there are positive technological solutions focused on sustainable battery production including direct lithium extraction (DLE) or battery repurposing and recycling. Nonetheless, these technologies together are not enough to diminish the scale of the transition or its carbon footprint intensity at the moment.

There are four primary forces which underpin this paradox for the battery metals industry (to be discussed in depth further down in this chapter): geopolitics, economics, technology, and legislation. These forces are intertwined and magnify each other in attempts to achieve a just green energy transition. As the nexus between national security and economic security continues to blur, the supply chains and business models of the past will need to be reconfigured at record pace and will cost hundreds of billions of dollars. The need for speed and capital deployment mean that larger amounts of fossil fuels and nuclear power will be required to accelerate this transition, as well as various battery metals projects, all scaling in tandem and on budget – a feat not typically achieved in the mining industry. In the near term, the costs of decarbonization could increase, owing to current multi-decade high inflation across the globe and underinvestment in battery metals capacity.

Moreover, market consensus for many battery metals calls for steadily increasing demand with a supply response that is likely to keep pace only in fits and starts. Products like lithium hydroxide or neodymium oxide are highly engineered specialty chemicals utilized in the battery supply chain and are expected to see strong growth prospects in the coming years (mining.com 2021). Despite the unique production process for these materials, they are not immune to the cyclical vagaries of the markets which, when coupled with their small market sizes, adds an additional layer of risk. Market forces can swing prices violently in one direction, rendering mining capacity in all but the lowest cost jurisdictions uneconomic.

When increased environmental, social, and governmental (ESG) scrutiny is factored in, the looming demand shock for battery metals produced in a more sustainable manner is certain to shift cost curves upwards, as costs for raw material production (and hence decarbonization) increase. In a competitive global market, the lowest cost producer wins, implying a race to the bottom. Given that the battery in an EV can account for 30% of the total cost of the car, managing costs is imperative for sustained profitability. If companies cannot pass higher costs down the supply chain, who will assume these increased costs? This is an important, and largely unaddressed question.

6.2 The Scale of Battery Metals Demand for Green Growth

While demand for battery raw materials will push higher, the actual trajectory will vary across metals including lithium, cobalt, graphite, copper, nickel, manganese, aluminum, and rare earth elements. Each of these resources has its own supply, demand, and pricing dynamic driven by a multitude of different end products, underpinned by high purity standards. That said, the lithium-ion battery and permanent magnet businesses are the true drivers of demand for these materials well into the future. Any decarbonization strategy must acknowledge this and find ways, perhaps through leveraging technology, to help supply grow from dozens of new mines in line with increasingly stringent ESG guidelines.

While all raw materials are vital to battery production, the most widely followed battery metal is lithium. Currently 550,000 tonnes per year (tpy) in size in terms of lithium carbonate equivalent (LCE) units, demand at roughly a 20% compound annual growth rate (CAGR) means that the lithium market will grow to over 1,100,000 tpy LCE by 2025 and over 2,800,000 tpy LCE by 2030. Current demand growth rates for lithium imply additional production capacity of five new lithium mines each year assuming a new lithium mine produces 30,000 tpy out to 2025. Additionally, fully integrated lithium production capacity can carry a capital expenditure of anywhere from US $500 million to US $1 billion. Capital availability is not the challenge here, but rather deploying such a vast amount efficiently is. This need to scale production in such a short period of time is unprecedented (Albemarle Corp 2021, 59).

As an example of battery pricing sensitivity, at spot lithium pricing of over US $70,000 per tonne in China as of September 2022, the cost of lithium in a 60 KWh battery with a nickel cobalt manganese (NCM) 523 chemistry amounts to roughly US $3360. While this is not terribly significant given the overall cost of the EV, using a spot lithium price from late 2020 during the depths of the COVID-19 pandemic of US $6500 for the same battery delivers a cost of US $312, a difference of 976%. The 2022 cycle peaks for copper, nickel, and aluminum, due to the re-opening of the global economy and speculative fever, demonstrate the potential effects high input costs can have on the battery supply chain. The elevated prices of West Texas Intermediate crude oil and natural gas in early 2022, due primarily to the fallout from the Russian invasion of Ukraine, also raised concerns as inflationary headwinds were persistent prompting a raging debate over transitory versus structural higher prices in the global economy. Permanently higher input costs emphasize the need to drive down battery costs through creative raw material sourcing and scale of battery factory plant buildout.

Markets such as copper or nickel, given their size, may be less exposed to such severe demand shocks and are unlikely to be bottlenecks to growth. The pricing transparency of both markets could help minimize any structural supply shortages before they emerge. Each of these raw materials, regardless of their market size are increasingly strategic in nature (Nassar and Fortier 2021). Without a reliable and steady supply of high purity battery materials, lithium-ion battery cell shortages are increasingly likely, and climate targets could be imperiled. It is perhaps best to think of the potential for bottlenecks along the battery metals supply chain on a "rolling" basis to 2030 with a metal such as lithium currently experiencing a structural shortage and other metals such as copper experiencing a similar dynamic later this decade.

6.3 Lithium-Ion Battery Economics: There Is No Free Lunch

The steady fall of lithium-ion battery costs has been reliable for decades. Since lithium-ion batteries were first introduced commercially in the 1990s, costs have fallen by as much as 97% owing to investments in research and development and scaling of battery factory output (Ziegler et al. 2021). These cost declines have been taken for granted in that it is assumed that they will continue indefinitely. Higher input costs due to economy-wide inflation have currently caused the price decrease of a lithium-ion battery to plateau, forcing industry participants to pause and consider how to operate in a higher cost environment. Trade wars, pandemics, and hot wars have in different ways informed government responses to faltering economic growth by flooding markets with liquidity. This stimulus has been delivered to maintain employment and economic growth in the name of social stability. The unprecedented volume of liquidity and delivery mechanisms (PPP loans, quantitative easing) predictably caused a persistent inflationary wave throughout 2022 reminiscent of the 1970s. Global central banks have begun to contain inflation through an aggressive interest rate tightening cycle and removing of liquidity from the markets through a process known as "quantitative tightening". Inflation has skyrocketed throughout the global economy and the lithium-ion battery business is not immune to this pervasive dynamic.

Battery grade lithium carbonate, one of two standard forms of lithium utilized in the battery business, experienced a price increase from roughly US $6500 per tonne in mid-2020 to over US $70,000 per tonne in 2022 on the spot market in China according to FastMarkets. Even long-term contract prices between mining companies and their customers saw rampant price spikes around US $30,000 to US $40,000 per tonne. While this may be good for the bottom line of lithium mining companies, it is decidedly bad for the rest of the supply chain that is now forced to pay higher pricing for the most essential of all battery raw materials. While other battery raw materials such as nickel, copper, cobalt, and rare earth elements all saw price spikes as the global economy reopened throughout 2021, many have seen prices revert to lower levels in 2022 due to factors including China's Zero COVID policy and fears of a global recession.

Battery metals were not the only commodities to experience extreme pricing volatility. Oil and natural gas saw similar price spikes and remained at elevated levels as well throughout 2022. These elevated input costs may have structural implications for the lithium-ion battery business if scale cannot outrun these higher costs. It remains to be seen how companies focused across the lithium-ion supply chain will pass on increased costs to their customers (or whether they will just absorb the costs and erode their profit margins in an increasingly competitive business). The automotive OEMs were among the first to respond through a series of price increases for their EV offerings. Throughout the Summer of 2022, General Motors, BYD, Tesla, and Rivian all announced price increases due to inflation from elevated input costs (CNBC 2022).

6.4 Lithium-Ion Battery Geopolitics: A Three-Way Race to Self-Sufficiency

The lithium-ion battery supply chain is fragmented, owing to the relative geologic concentration of raw materials in countries such as Chile, Australia, or Russia, the concentration of midstream processing capacity in China, and the concentration of end markets mainly in Asia. The three exogenous shocks mentioned earlier (the US/China trade war, COVID, and the invasion of Ukraine) have drawn the attention of policymakers and business leaders toward a policy that minimizes the complexity of these supply chains by relocating or rebuilding as much infrastructure as possible closer to home. The notion of "friendshoring" is built on the simple premise that a lithium-ion supply chain should not touch China and must have complete built in visibility to satisfy an increasingly stringent ESG lobby. The thinking has been that rebuilding supply chains to align with the "friendshoring" philosophy satisfies both national and economic security, minimizes dependency on strategic adversaries such as China and Russia, and creates new trading systems between the United States and its allies. However, rebuilding supply chains and making them more streamlined will increase the cost for all involved, not to mention create competition for resources between friend and foe alike.

If the goal is to rebuild (or initially build) lithium-ion battery infrastructure including manufacturing plants and sourcing materials locally, it stands to reason that costs along the supply chain will increase. This will force lithium-ion supply chain participants such as auto OEMs and battery manufacturers to choose how much of these increased costs they are willing to internalize versus how much they will pass on to their customers. Just how willing customers are to buy premium priced EVs remains to be seen, though the data is encouraging given Tesla's success in selling EVs to the upper income segment of Western markets and in China. Of course, successful EV penetration will hinge on the ability of middle-class customers to afford this mode of transportation and it is likely that additional policy tools may be needed should raw material prices stay elevated for longer than assumed.

As the success of any new supply chain rests on the sourcing of the raw materials used, this may result in new hyper-competition for access to the raw materials. Given China's historical dominance of the lithium-ion supply chain, many had viewed supply chain dynamics as a race to the bottom for lowest cost production and materials. This thinking played a significant (but not exclusive) role in these supply chains settling in Asia and importantly in China. Now that geopolitics have clearly dictated a rethink of traditional globalization, China, the US, and the EU, as three of the largest consumer markets in the world, will be competing for both natural resources and intellectual know-how to build the resilient supply chains of the twenty-first century, implying higher prices across reconfigured supply chains. This will also highlight areas where China is dominant such as raw material refining and purification.

The German Chancellor Olaf Scholz' August 2022 visit to Canada, accompanied by Mercedes-Benz and Volkswagen, that led to the signing of memoranda of understanding between these companies and the Canadian government to deepen cooperation around raw material access and processing, offers a telling example. This dovetails on a strategic partnership between the EU and Canada commenced in June of 2021 (European Commission 2021) focused on raw materials and the cross-border integration of supply chains. This partnership shines a light on how the EU intends to foster the growth of clean tech supply chains while at the same time encouraging the security of supply of raw materials.

The Russian invasion of Ukraine has also highlighted the paradox of green growth in that while Russia is a major producer of Class 1 nickel, required for use in certain lithium-ion battery cathodes, sanctions have rendered difficult exporting material into Western markets. It is likely that this material will end up in China, bolstering the country's supply chains. This is not to say that Western sanctions should not be in place, but it does demonstrate that in a globalized economic system, valuable materials will find a home regardless of the geopolitical backdrop.

Indonesia, moreover, is another example of where resource nationalism has reared its head. The country is and will remain a major producer of nickel for the energy transition. However, given the increasingly technical processes such as high-pressure acid leach (HPAL) to produce the material and the high levels of production that will be powered by coal, raise questions about how "green" this metal truly is. Elon Musk's promise of a "giant contract for a long period of time" (Mining-Technology.com 2020) for sustainably produced nickel will likely end up going through Indonesia for lack of near-term, lower carbon scalable alternatives. Furthermore, China's dominant presence in Indonesia as both a miner and producer of nickel products has only served to underscore how truly strategic Indonesian officials think nickel is and will become in the future.

Based on this reality, it has not been surprising to see automotive manufacturers trying to circumvent geopolitical issues by moving towards a vertical integration strategy to protect margins and better control their supply chains. Chinese companies have for some time employed this strategy, however, a recent example in North America is indicative of how the tide is now changing in the West. General Motors has embarked on a strategy to source lithium, produce its own battery cells, and recycle battery scrap and spent EV batteries all in the United States. While this makes strategic sense for the company, the race for resources, intellectual talent, and customers indicates that the green transition will be costly for GM given that it faces growing competition from both allies and adversaries alike.

6.5 Lithium-Ion Battery Legislation: Governments Get in the Game

Many investors look at how China has come to dominate the lithium-ion supply chain and wonder if Western governments will ever be able to compete. While the United States and China are playing the same game (green growth through industrial policy), the two countries are competing based on a different set of rules. The United States relies mostly on the financial markets for investment capital, for instance, while China employs a more top-down state-led investment structure. The growing geopolitical contention, however, has spotlighted the differences in domestic legislative realities (Kalantzakos 2022). China's state-directed capitalism enjoys some support mainly because of its ability to speedily react to external developments. In the United States, the idea that government may need to become more involved in funding lithium-ion supply chain projects and effectively "pick winners" remains problematic especially amongst Republicans. That being said, and while distrust of government remains high, the Biden administration has taken unprecedented and unapologetic steps in support of the green transition, as indicated by the passing for the Inflation Reduction Act (United States Congress 2022b).

Financial markets alone, however, are hesitant to fund an upstream green transition given raw materials pricing volatility, small markets, boom and bust investment cycles, and supply chain concentration. This has prompted Western governments to think creatively about providing funding across the supply chain while not acting as an explicit backstop for various projects. There are already examples of public/private partnerships such as the Department of Energy AVTM loan to Tesla for US $465M working as intended, but these types of loans, grants, and tax breaks must be focused on the upstream portion of the lithium-ion supply chain to speed the build out of a more localized supply chain structure.

A good example is Title III of the Defense Production Act (DPA) in the United States. This Korean War-era law "authorizes and enables the Department of Defense to undertake actions, including but not limited to, feasibility studies and modernization projects for mature mining, beneficiation, and value-added processing projects to increase productivity, environmental sustainability, and workforce safety" (United States Department of Defense 2022). In recent years, this act was invoked by both President Donald Trump and President Joe Biden based on the realization that as economic and national security become more intertwined, a coherent government policy on the sustainable and sufficient sourcing of critical raw materials is vital for economic growth. Efforts such as this must be redoubled along with support for loan programs administered by the Department of Energy.

While it is early days, it remains to be seen how effective these programs are at spurring broad domestic lithium-ion supply chain development. The underinvestment in the lithium-ion supply chain coupled with underinvestment in fossil fuel infrastructure in recent years has served to raise costs along the supply chain and highlights a central truth of supply chain self-sufficiency: there is no resilient downstream (battery or EV production) without a resilient upstream (domestic mining and refining of battery metals).

Fortunately, there is legislative momentum in the West. Under President Biden, the Bipartisan Infrastructure Act and most recently the Inflation Reduction Act are aimed at bolstering a domestic clean energy ecosystem, and the lithium-ion battery figures prominently as previously discussed. The Bipartisan Infrastructure Act will allocate more than US $7 billion in the battery supply chain in the United States

(United States Department of Energy 2021). The Inflation Reduction Act takes this one step further by helping to support the construction of a full domestic lithium-ion supply chain through an extension of tax credits but also requiring minimum percentages of domestic content that must be extracted, processed, or recycled in North America or countries that have a Free Trade Agreement (FTA) in place with the United States. This commences with a 40% minimum domestic content requirement and escalates at 10% per year throughout this decade. Given that there is only one operating lithium mine currently in the United States forecast to produce up to 10,000 tonnes per year of material and no other domestic lithium mines expected before 2025, the United States EV market will be reliant on countries including Chile or Australia for its immediate lithium raw material needs.

Also, of note, Indonesia, whose nickel supply for the lithium-ion battery industry is central as previously discussed, does not have an FTA in place with the United States. So, although the definition of "domestically sourced" raw material is intentionally wide, this doesn't address the need to lower the overall carbon footprint of lithium along the supply chain given increased transportation and logistics needs and further highlights the paradox of green growth. Emissions along the lithium-ion supply chain are, therefore, likely to rise before they fall in the future under these rules. Ironically, one measure that could expedite the buildout of domestic mining capacity is not addressed in this legislation and that is permitting reform. While there are reports that the US Congress will address this glaring inconsistency in the future, the most recent attempt to do so by Senator Joe Manchin was pulled from broader legislation as it lacked the adequate number of votes to be enshrined into law (Bloomberg Law 2022).

The European Union has chosen a similar approach, with a mix of both incentives and regulations designed to encourage domestic supply chain self-sufficiency (European Battery Alliance 2022). A key difference concerns the reuse and recycling of lithium-ion batteries. By 2030, any recycler of lithium-ion batteries must demonstrate a minimum recovery rate of critical battery materials including lithium carbonate, nickel sulfate, and cobalt sulfate and these thresholds will increase by 2035. Additionally, for those same companies, by 2030 they must certify that a minimum percentage of recycled content is used in new lithium-ion batteries. These percentages will also increase by 2035. Standards such as these can help spur innovation along the supply chain.

While the United States has a longstanding tradition of letting the free market sort out supply chain issues, China's dominance over many aspects of the supply chain and the collective need for cross-industry decarbonization likely require some form of government involvement. According to Bloomberg New Energy Finance, China controls 80% of battery raw material refining, 77% of battery cell production, and 60% of component manufacturing (Bloomberg New Energy Finance 2022). While this level of dominance did not happen overnight, it did occur thanks in part to China's economic model of state supported capitalism. It remains to be seen how the new legislation in the United States and European Union blunts this Chinese dominance and in what timeframe.

6.6 Lithium-Ion Battery Technology: Panacea or Not?

Societal advance through technological innovation has been a powerful growth engine for centuries. The lithium-ion battery, a technology first commercialized in the early 1990s, is one important example that has enabled the growth of other technologies such as handheld electronics, grid scale energy storage, and EVs. An increasingly globalized economy has allowed countries to access and exploit these technologies to enhance their citizen's standard of living. This has created resource dependencies and an expectation of ever-increasing standards of living, underlining the importance of flexible and short supply chains in contrast to what exists today.

As the effects of climate change, however, are felt with greater frequency, whether it be heatwaves in China, floods in Pakistan, or dried riverbeds in Europe, business as usual approaches are no longer viable. With the urgency for decarbonization clear, it is important to realize that achieving success, however one defines it, will require billions of dollars in capital deployed across different technologies including batteries, carbon capture, hydrogen, and nuclear power. As the saying goes, "nothing cures high prices, like high prices". However, perhaps the phrase should read "nothing cures high prices like technology and innovation." Due to these imperatives (high prices, carbon emissions, and security of supply), investors and stakeholders have begun to devote an increasing amount of attention to two technologies that can help blunt the sting of high prices and resource dependency: lithium-ion battery recycling and direct lithium extraction.

6.7 Direct Lithium Extraction

One possibility for sustainable growth in lithium supply involves direct lithium extraction (DLE) technologies. These include ion exchange, solvent extraction, and adsorption or some combination thereof. These technologies are not new and are utilized for lithium extraction by Livent at their Salar de Hombre Muerto in Argentina and Chinese lithium producers in Qinghai province in China. These processes are similar in that they each use the unique physical properties of a solid or liquid material to separate and selectively extract lithium from brine while ignoring other elements such as potassium or magnesium. There has been an increased amount of interest in DLE in recent years because of the desire for the investment community to allocate capital towards ESG-friendly technologies. Lithium production from brines can put pressure on freshwater aquifers, is capital intensive (pond lining can account for 40% of project capital expenditure), reagent intensive (chemicals can account for 60% of project operational expenditures), time intensive (brine processing can take from 12 to 18 months to produce lithium carbonate) and has low recovery rates. Finding a magic bullet that can solve for these issues has long been a goal of the lithium industry and has become more acute as lithium demand has accelerated.

While promising, DLE has yet to grow into the panacea that the lithium industry has hoped for. The technologies are not new or novel and may arguably offer a lower capital and carbon intensive route to lithium production. However, much of the technology involved has yet to truly scale to the point where adequate investment dollars can generate an acceptable return to compensate for the heightened level of technology risk. In addition to this, DLE processing technology could use as much if not more freshwater when compared to a traditional lithium brine operation, again raising the paradox of a "green" lithium extraction technology being just as resource intensive as traditional mining methods.

Based on author estimates, should all current DLE projects that have stated their intentions to achieve commercial production do so by 2030, DLE-sourced lithium could account for as much as 15% of total lithium production. DLE should be thought of as additive to existing lithium processing. It holds promise as a route to minimizing carbon emissions in lithium brine production, but the economics will still need to be better understood. As of mid-2022, roughly US $1 billion has been raised or committed to develop and perfect the technology. Based on publicly available cost information from the National Renewable Energy Lab (NREL) (Warren 2021) the operating expenditure for a producing lithium project utilizing DLE technology could range between US $3,500 and US $4,500 per tonne. For perspective, this would place a DLE project near the bottom of the global lithium cost curve, in between traditional lithium brine projects and traditional hard rock lithium mines. While the economics are promising, the ability to prove that the technology is more efficient at scale remains a challenge.

6.8 Recycling

Lithium-ion battery recycling has gained a renewed focus as a source of raw material supply. Recycling of lithium-ion batteries has historically suffered from poor economics based on low to marginal resource recovery rates and low metals prices. With lithium alone proving too costly to recycle and refine to battery quality, lithium-ion battery recycling has traditionally been driven by the value associated with cobalt and nickel. While these metals are vital to sustainable recycling economics, this is changing with advances in hydrometallurgical processing techniques which have a lower energy intensity relative to other recycling techniques such as pyrometallurgy. State-of-the-art hydrometallurgical recycling processes (or hydrometallurgy combined with pyrometallurgy) can yield recovery rates over 90% for most materials (Bruckner et al. 2020).

As EV sales continue to ramp during this decade, the main feed for recyclers will be scrap from batteries as well as small format batteries originating from cell phones or laptops. Larger format EV batteries are not likely to be a significant source of feed for recycling until into the 2030s given that a large format EV battery can have a useful life of at least 12 years. With scores of Gigafactory-scale battery manufacturing plants in production, under construction or planned to 2030 amounting to over 6 TWh of capacity according to Benchmark Mineral Intelligence (Benchmark Mineral Intelligence 2022), the amount of battery scrap produced will provide a steady stream for the existing recycling business both inside and outside China. With reject rates in new battery plants running as high as 30% during the initial year or years of production, battery scrap can serve as a partial buffer against any potential demand shocks in battery metals during this decade. While battery scrap is not uniform in nature, any recovered lithium carbonate, cobalt sulphate, or nickel sulphate would be welcome by battery manufacturers intent on a strategy of vertically integrating their supply chain.

Much of the industry today is focused on hydrometallurgical processing, with its lower operating costs and lower carbon footprint relative to pyrometallurgical processing which involves the smelting of the battery. The higher energy requirements are the main factor behind the higher operating expense of pyrometallurgy, so plant location and access to affordable energy is of paramount importance. Regardless of whether traditional hydrometallurgical or pyrometallurgical processing routes are pursued, the lithium in a lithium-ion battery has traditionally been difficult to extract and purify to battery grade. Recycled lithium from batteries was typically destined for the construction business in cement owing to the prohibitive cost of upgrading lithium chemicals to battery quality. With long term lithium pricing forecast to settle higher relative to historic norms, this may help to increase the amount of recycled lithium available within the lithium-ion ecosystem.

6.9 A Note on Substitution

Another solution to natural resource shortages is substitution. There is a great deal of focus on research and development of battery chemistries which are designed to provide the optimal blend of safety and performance while using fewer materials and lowering the carbon footprint of the production process. In almost every case, however, minimizing or substituting one raw material means adding more of another material. The evolution of the NMC lithium-ion battery chemistry is an instructive case study. As the EV industry moves to address range anxiety and diminish worries over how cobalt is sourced and mined, the most popular lithium-ion chemistry for superior energy density is evolving towards NMC 811 (approximately 80% nickel, 10% manganese, 10% cobalt). While there will not be a single battery chemistry that is utilized across platforms, the industry shift to "nickel heavy" and "cobalt light" cathode materials is well on its way. Separately, lithium iron phosphate (LFP) batteries have continued to maintain and grow market share due in part to concerns over nickel and cobalt availability. While this may help to alleviate dependencies on critical materials somewhat, minimizing the cobalt content and increasing nickel content can present new challenges with increased instability of the battery (Ding et al. 2019). This means that existing battery chemistry formats that have a longer history of use in different environments such as LFP may be in use longer than first thought.

Aside from manipulating existing chemistries, there exists the possibility of using alternative battery chemistries altogether. A current example is an increasingly loud chorus of industry participants pointing towards sodium-ion batteries as a substitute for lithium-ion in mobility. CATL, one of the largest lithium-ion battery manufacturers in the world, is the most visible leader in this regard (CATL 2021). There are limitations in the technical abilities of sodium-ion relative to lithium-ion, but should raw material pricing remain elevated, battery technologies such as sodium-ion, which offer an acceptable blend of cost and performance could increase market share in the coming years.

6.10 Conclusion

Climate change is not a threat to be dealt with in the future, but rather a threat to be dealt with today. Recent legislation in the United States and other parts of the world indicates that governments are serious about this threat and have begun to forge a pathway to sustainable decarbonization by promoting the domestic restructuring of clean energy supply chains. To achieve substantial lithium-ion battery production scale domestically by 2030 and beyond, enormous amounts of mined raw material will be required. This is the key lynchpin in the execution of global green industrial policy. As discussed above, battery recycling and DLE are not likely to make a meaningful dent in sustainable battery metals supply before the mid-2030s. In the interim, all battery metals markets will need to scale flawlessly and aggressively.

This will be no easy feat when one considers that these are specialty chemicals produced to an exacting and evolving purity while consumers demonstrate increasing concern about the ESG footprint of entire supply chains. For all the optimism surrounding the green transition, this new supply will come at a financial and environmental cost. Raising and allocating the hundreds of billions of dollars in capital needed to expand production capacities for battery metals including lithium, cobalt, copper, rare earth elements, or nickel is not the most pressing challenge in the race to decarbonize. Much more challenging is allocating that capital quickly enough to meet global decarbonization goals with a focus on sustainability.

It must be acknowledged, moreover, that initially more, not less, fossil fuel will be required to build a greener economy and associated supply chains. Moreover, successful execution of a regionalized supply chain rebuild will require consensus and pragmatic decision-making between governments and the private sector around the best way to scale and decarbonize at the same time. The demand for battery metals appears increasingly robust, so a coordinated effort between the public and private sectors which focuses on R&D and project development is the optimal way to ensure supply from domestic sources or other allies. Put another way, while mechanisms such as carbon taxes may be a preferred method to encourage decarbonization, the carrot and stick method mentioned earlier is likely the path of least resistance. The Biden Administration's actions indicate that it would appear to agree given the approach to vehicle electrification embedded in the Inflation Reduction

Act. A rapidly evolving technology landscape underpinning decarbonization including falling costs of lithium-ion batteries is likely to help, but efforts into battery R&D and other storage technologies must be redoubled. Governments picking winners is not a recipe for long-term success, but governments funding R&D coupled with tax credits and other investment incentives and letting market participants innovate from there certainly is.

The bullish narrative around battery metals has been amazingly resilient and has accelerated because of the effect of COVID-19 on supply chains. This perceived freight train of demand is about to run headlong into several industry-wide challenges including increasing product purity demands, ESG requirements, and an inflationary supply chain rebuild. A key takeaway is that cost curves for battery metals including lithium and cobalt will almost certainly have to rise. This ought to act as a floor for battery metals prices to remain higher for longer, incentivizing investment and increased supply over the long run. Battery metals producers that can finance, build capacity, and sell product to a growing list of downstream customers should benefit from the advantages of scaled production despite higher costs associated with inputs from fossil fuels.

In the twentieth century, owning the raw materials in a supply chain conferred a competitive advantage. Just ask Henry Ford. However, the twenty-first century demands something more, as raw material control is not enough and the need to decarbonize globalized industries is clear. Developing and owning the intellectual property around battery metals supply chains to ensure these supply chains are built at a minimum carbon and capital intensity is important. More important, however, is a society-wide pragmatic approach to lithium-ion supply chain development in the coming decades. This aspect of the global energy transition rests on increased supply of battery metals and the fossil fuels needed to bring them to market. As Milton Friedman once famously said, "there is no such thing as a free lunch (Friedman 1975)." As the global economy pushes towards an electrified future, these words are especially pertinent.

References

Albemarle.com. 2021. https://investors.albemarle.com/static-files/7b16ee83-eef0-4db9-a724-920e7df2ca5c. Accessed 30 September 2021.

Benchmark Mineral Intelligence. 2022. Lithium ion battery capacity pipeline breaks the 6 TWH capacity barrier. https://www.benchmarkminerals.com/membership/lithium-ion-battery-pipeline-breaks-the-6twh-barrier/. Accessed 4 September 2022.

Bruckner, L., J. Frank, and T. Elwert. 2020. *Industrial recycling of lithium-ion batteries – A critical review of metallurgical process routes.* https://www.mdpi.com/2075-4701/10/8/1107/pdf. Accessed 11 October 2021.

Bloomberg New Energy Finance. (2020). https://about.bnef.com/blog/china-dominates-the-lithium-ion-battery-supply-chain-but-europeis-on-the-rise/. Accessed 4 September 2022.

Bloomberglaw.com. Lee, S. 2022. Manchin's agreement paves way for permitting changes in fall. https://news.bloomberglaw.com/environment-and-energy/manchins-agreement-paves-way-for-permitting-changes-in-fall. Accessed 3 September 2022.

CATL. 2021. CATL unveils its latest breakthrough technology by releasing its first generation of sodium-ion batteries. https://www.catl.com/en/news/665.html. Accessed 2 September 2022.

CNBC.com. Rosevear, J., and M. Wayland. 2022. Automakers are jacking up prices on electric vehicles to bake in rising materials costs. https://www.cnbc.com/2022/05/21/electric-vehicle-prices-rise-at-tesla-rivian-gm-and-other-carmakers.htm. Accessed 2 September 2022.

Ding, Y., Z. Cano, A. Yu, J. Lu, and Z. Chen. 2019. *Automotive lithium-ion batteries: Current status and future perspectives*. https://www.osti.gov/pages/servlets/purl/1561559. Accessed 11 October 2021.

European Battery Alliance. 2022. EU legislation and directives. https://www.eba250.com/legislation-market/eu-legislation/. Accessed 3 September 2022.

European Commission. 2020. CO2 emission performance standards for cars and vans. https://climate.ec.europa.eu/eu-action/transport-emissions/road-transport-reducing-co2-emissions-vehicles/co2-emission-performance-standards-cars-and-vans_en. Accessed 6 November 2022.

———. 2021. EU and Canada Set Up A Strategic partnership on raw materials. https://single-market-economy.ec.europa.eu/news/eu-and-canada-set-strategic-partnership-raw-materials-2021-06-21_en. Accessed 7 November 2022.

Friedman, M. 1975. *There is no such thing as a free lunch*. Chicago: Open Court Publishing Company.

Grant, A., D. Deak, and R. Pell. 2020. The CO2 impact of the 2020's battery quality lithium hydroxide supply chain. https://www.jadecove.com/research/liohco2impact. Accessed 11 October 2021.

International Energy Agency. 2022. Global electric car sales have continued their strong growth in 2022 after breaking records last year. https://www.iea.org/news/global-electric-car-sales-have-continued-their-strong-growth-in-2022-after-breaking-records-last-year. Accessed 2 September 2022.

Kalantzakos, S. 2022. Interview with expert. September 2022. See Relevant chapter in the volume.

Mining-technology.com. Hall, M. 2020. Elon Musk pledges "giant contract" for sustainable nickel miners. https://www.mining-technology.com/analysis/elon-musk-pledges-giant-contract-for-sustainable-nickel-miners/. Accessed 3 September 2022.

Mining.com. 2021. Magnet rare earth oxides market to increase fivefold by 2030. https://www.mining.com/magnet-rare-earth-oxides-market-to-increase-fivefold-by-2030-report/#:~:text=A%20new%20report%20by%20Adamas,from%20the%20current%20%242.98%20billion. Accessed 11 October 2021.

Nassar, N., and S. Fortier. 2021. Methodology and technical input for the 2021 review and revision of the U.S. *Critical Minerals List*. https://pubs.usgs.gov/of/2021/1045/ofr20211045.pdf. Accessed 11 October 2021.

U.S. Department of Defense. 2022. Defense Production Act Title III presidential determination for critical materials in large-capacity batteries. https://www.defense.gov/News/Releases/Release/Article/2989973/defense-production-act-title-iii-presidential-determination-for-critical-materi/#:~:text=The%20DPA%20Title%20III%20authorities,environmental%20sustainability%2C%20and%20workforce%20safety. Accessed 3 September 2022.

United States Congress. 2022a. H.R. 5376 – Inflation Reduction Act of 2022. https://www.congress.gov/bill/117th-congress/house-bill/5376/text. Accessed 6 November 2022.

———. 2022b. H.R. 5376 – Inflation Reduction Act of 2022. https://www.congress.gov/bill/117th-congress/house-bill/5376/text. Accessed 7 November 2022.

US Department of Energy. 2021. DOE fact sheet: The Bipartisan infrastructure deal will deliver for American Workers, families and usher in the clean energy future. https://www.energy.gov/articles/doe-fact-sheet-bipartisan-infrastructure-deal-will-deliver-american-workers-families-and-0. Accessed 3 September 2022.

Walton, B., J. Hamilton, and G. Alberts. 2020. Electric vehicles – Setting a course for 2030. https://www2.deloitte.com/us/en/insights/focus/future-of-mobility/electric-vehicle-trends-2030.html. Accessed 11 August 2021.

Warren, I. 2021. *Techno-economic analysis of lithium extraction from geothermal brines*. Golden: National Renewable Energy Laboratory. NREL/TP-5700-79178. https://www.nrel.gov/docs/fy21osti/79178.pdf. Accessed 11 October 2021.

Ziegler, M., J. Song, and J. Trancik. 2021. Determinants of lithium-ion battery cost decline. *Energy & Environmental Science* 14: 6074. https://pubs.rsc.org/en/content/articlepdf/2021/ee/d1ee01313k. Accessed 2 September 2022.

Chris Berry is president and founder of House Mountain Partners, LLC and is based in Washington DC. Chris Berry has been an independent analyst for twelve years with a focus on battery metals supply chains including lithium, cobalt, graphite, nickel, copper, vanadium, and rare earths. His work supports a roster of global clients including investors; corporates including mining, oil and gas, and electric vehicle producers; and academia. He has visited and performed economic analysis on mines, refineries, and other businesses along the battery metals supply chain across six continents. His advisory work has a specific focus on how battery metals supply chains are evolving to create opportunities and threats during the energy transition in the coming decades. Aside from traditional supply, demand, and pricing analysis, his recent work has focused on battery recycling, alternative extraction technologies, and how battery metals will play an increasingly strategic role in geopolitical and supply chain realignment. Before shifting focus to analysis of these trends, Chris gained twelve years of capital markets experience in New York. Chris holds a Master of Business Administration in finance with an international focus from Fordham University and a Bachelor of Arts in international studies from the Virginia Military Institute.

Chapter 7
Raw Material Demands for the Green Transition: Risks, Opportunities, and Required Actions to Meet the 2030 Climate Targets

Roland Gauß, Carsten Gellermann, and Alexander Maurer

Abstract Renewables, electric mobility and the building of a circular bio and hydrogen economy are at the heart of Europe's green transition. The EU however, like other nations, has underestimated the material intensity of the green transition including the unprecedented levels of extraction it generates, the socio-economic impacts along the entire life cycles of technologies and related materials value chains, and its significant global environmental footprint. To fulfill its ambitious climate neutrality goals in competitive market conditions, nonetheless, the Union must ensure a sustainable supply of raw materials and advanced materials, ramp up its industrial capacity for a circular economy, and ensure a resilient supply of low-cost clean energy. Yet Europe remains highly import dependent on critical and strategic raw materials, which include, for example, silicon, copper, rare earths, nickel, cobalt, graphite, manganese, and platinum group elements. Moreover, parts of the value chain are highly concentrated, often in China. This chapter will examine how the EU, and Germany in particular, now address these important shortages and challenges by recognizing the need for joint public private investments to not only shoulder the green transition but also to support new market opportunities. The Union has, thus, created industrial alliances on hydrogen, batteries, semiconductors, and raw materials to address regulatory bottlenecks and spotlight investment oppor-

R. Gauß (✉)
EIT RawMaterials, Berlin, Germany
e-mail: roland.gauss@eitrawmaterials.eu

C. Gellermann
Fraunhofer Institute for Silicate Research ISC and Fraunhofer Network Nanotechnology, Wuerzburg, Germany
e-mail: carsten.gellermann@isc.fraunhofer.de

A. Maurer
AlexMConsulting, Mannheim, Germany
e-mail: alexandermaurerrheinau@gmail.com

S. Kalantzakos (ed.), *Critical Minerals, the Climate Crisis and the Tech Imperium*, Archimedes 65, https://doi.org/10.1007/978-3-031-25577-9_7

tunities. To enable the green transition, however, further support is required for the creation of robust circular value chains with global as well as domestic footprints.

Keywords Green transition · Critical raw materials · Circular economy · Hydrogen · Bio economy

7.1 The Challenge

Nations worldwide have announced ambitious goals to significantly reduce their CO_2 footprints and have defined quantitative targets to achieve climate neutrality (United Nations 2015). The EU, more specifically, declared that it will reduce its greenhouse gas emissions to at least 55% of its 1990 levels by 2030, and will become climate neutral by 2050 (Council of the European Union 2019; European Commission 2021). The latest report of the Intergovernmental Panel on Climate Change (2021) underlines the relevance and urgency of achieving these targets. Currently, in the EU a consensus has been achieved on ways to innovate on how energy is used, how the transport of people and goods is organized, and how sustainable industrial production can be promoted. However, the challenges and opportunities regarding the raw materials dimension of this disruptive change are still massively underestimated. The expression "climate neutrality" can be misleading because the Green Transition fundamentally depends on the exponential growth in demand for raw materials, i.e., primarily metals, minerals, and advanced materials as well as biomass. The production of these resources comes with extensive environmental footprints.

In Europe, political decision makers and industry envision, for example, a hydrogen based industrial economy, which would be instrumental to produce products like "green steel" and to enable new forms of "Zero Emission Mobility". The success of the hydrogen transition however, from both an economic and sustainable perspective, depends on the rapid deployment of renewable resources at competitive prices in order to produce hydrogen at scale. Unfortunately, energy prices have skyrocketed and remain volatile, a trend reinforced by the Russian invasion of Ukraine, making this transition further costly for society. Simultaneously, the renewable energy sector is struggling to make profit and expand the infrastructures required, particularly due to slow permitting procedures for new installations (WindEurope 2022). In addition, raw materials (the prices of which have been also skyrocketing) are a key cost driver in renewable energy and significantly determine the environmental footprint of the technologies (Agora Verkehrswende 2019). Moreover, the extraction and processing of many of the raw materials and advanced materials that are needed for their production are highly concentrated in a few countries resulting in supply chains with low resilience. Furthermore, there is a lack of standards and transparency in determining the specific provenance of materials and the social and environmental conditions of their production.

More importantly, there is a real danger that the discourse and decision-making process of climate neutrality targets will be reduced to a consideration of technological choices in downstream manufacturing industry and energy sectors without thoroughly addressing the consequences in terms of costs, environmental footprints, and societal aspects along the entire life cycles of technologies and related materials value chains. Moreover, as Europe's import dependency on raw materials and advanced materials has increased, problems such as waste and CO_2 emissions linked to raw materials extraction and processing have been exported to other parts in the world.

This chapter examines how the EU, and Germany in particular, needs to address the changing demand scenarios of supply chains required to scale up sustainable mobility and a bio- and hydrogen-based economy in the context of the aforementioned challenges. It has been argued that the green transition comes not only with fundamental challenges and risks but also with significant opportunities. The Union has, thus, created industrial alliances on hydrogen, batteries, semiconductors, and raw materials to address regulatory bottlenecks and spotlight investment opportunities. Among these alliances, the European Battery Alliance and the European Raw Materials Alliance have provided direction and delivered results that are helping the EU build resilience and diversify supply chains to mitigate risk. To enable a successful green transition that will make Europe the first climate neutral continent by 2050, however, further support is required for the creation of robust circular value chains with global as well as domestic footprints.

7.2 Sustainable Materials for Future Energy and Mobility

The energy and mobility transitions come with a disruptive change in raw materials and advanced material demands. With respect to energy, hydropower accounted for the largest share in the renewables sector, i.e., 17% of global energy generation (IEA 2021). Due to the emerging wind and solar power sectors, which – in fact – outpace all other renewable energy sources including geothermal and bioenergy (IEA 2021) however, hydropower's percentage share in the energy market is shrinking. Yet, the accelerated deployment of wind and solar energy comes with exponentially growing material demands. Forecasts in absolute numbers fluctuate based on the overall installation scenario as well as the specific technologies and materials used. Because of the intermittent nature and specific local availability of wind and solar energy, moreover, a greater need for energy conversion, storage, and transport is inevitable, which further increases material demands. While this poses a significant challenge, it also points to the huge potential of regaining materials from end-of-life wind and solar energy systems. Whereas fossil fuels are consumed in chemical energy conversion processes, functional materials in wind and solar as well as energy storage systems like magnetic materials, semiconductors, electrodes, and membranes can be re-used or recycled. Hence, if the energy and mobility transitions are designed to go hand in hand with an industrial capacity building for the circular economy, the challenge of growing material demand is not only addressed but provides an opportunity for substantial economic growth (Gregoir 2022).

7.2.1 Photovoltaics

According to the International Energy Agency (IEA 2021), global installation of photovoltaics will most likely be dominated by crystalline silicon-based technologies until 2030. Annual installations will grow from 100 GW today, to 150–280 GW worldwide by 2030, i.e., depending on the installation scenarios. This capacity growth would correspond to an enormous increase in the demand for various materials, including concrete, plastics, glass, steel, aluminum, copper, silicon, and silver. Global annual demand for copper and silicon, for example, may more than double to about 800,000 t each. Both materials have limited production in Europe. Whilst the global production of copper is still diversified to some degree, the production of silicon is highly dominated by China, which controls two thirds of global production. In addition, the manufacturing of silicon wafers and cells are dominated by China, which makes Europe completely import dependant along the entire value chain of this emerging technology. Recycling schemes have not yet been installed on a large industrial scale in Europe, which limits the possibility of alternative, less energy intensive supply. In fact, the major energy source for Chinese silicon and silicon wafer production is coal (European Commission 2020).

7.2.2 Wind Energy

A similar but different picture evolves around wind energy. Europe is still a global leader in the installation as well as the manufacturing of wind turbines, including key components like blades and generators. On a global level, anticipated growth is factored to scale between 90 and 150 GW annual added capacity by 2030, i.e., way above the 60 GW annual installations in 2019 (IEA 2021). In its recently announced Fit-for-55 industrial strategy update (European Commission 2021), the European Commission has announced its intention to further increase the Union's targets of installed wind energy capacity from 180 GW today to 451 GW by 2030. Thus, the EU alone will have to install 30 GW wind power capacity per year until 2030. So-called Direct Drive wind turbine systems installed offshore show a particularly high growth potential due to their high energy efficiency and robustness. This technology uses up to 650 kg of rare earth permanent magnets (Nd-Fe-B) per megawatt. Depending on the anticipated growth scenario, 15,000–20,000 t of Nd-Fe-B would need to be used for the EU wind energy sector alone, i.e., three to four times more than today (according to high demand scenario (HDS) by JRC 2020). However, the entire rare earth supply chain from mining and refinement as well as metal, alloy, and magnet making is dominated by China, and – at the time of writing – continues to lack transparency and recognized sustainability standards. More than 90% of sintered Nd-Fe-B is made in China (Gauß et al. 2021).

7.2.3 Electric Vehicles

Similarly, the electric vehicle (EV) sector benefits from using high energy efficiency, so-called synchronous electric motors that use permanent magnets to provide highest torque and to extend driving range as much as possible. Global rare earth magnet demand may rise from less than 5000 t to 40,000–70,000 t by 2030 (Fig. 7.1; Gauß et al. 2021). An even larger materials demand impact, in relative and absolute terms, is linked to the growing demand for battery materials, particularly those found in the anode and cathode, which may include graphite, silicon, nickel, cobalt, manganese, lithium, aluminium, and copper. Today, there are competing lithium-ion battery technologies. The raw materials demand per gigawatt hour energy storage will depend on the innovation scenario. As an example, battery grade nickel demand will grow from a few thousand tons to somewhere between 1,000,000 and 2,000,000 t (IEA 2021); thus, battery cell manufacturing will be the largest driver in the growth of nickel extraction and processing (DERA 2021). While in principle nickel is not considered to be a scarce metal, industrial capacities to produce *battery grade* nickel and respective cathode precursor materials will have to be put in place to meet the exponentially growing demand. In order for the EU's ambition to produce one third of global battery capacity by 2030 (Wirtschaftswoche 2019) to be realized, industry and political decision makers will have to find ways to sustainably secure these large amounts of materials, that is, all along the value chain. Responsible sourcing, sustainable materials manufacturing, and establishment of a circular economy of these materials should be considered the highest priority, because the footprints of the materials fundamentally determine the overall footprint of the batteries and those of the entire clean technology product. The disruptive change in the energy and mobility sectors and the exponentially growing demand for a new set of raw materials require – by definition – a major investment in primary raw materials extraction. Once the market will enter a state of saturation, urban mining can serve as a major source of raw materials (Fig. 7.2; Gregoir 2022).

For solar, wind and EV production, strategic investment in the upstream sector will not only help secure supply but will also ensure job creation. Electric vehicle battery production, more specifically, demonstrates that the job market related to the growing battery sector may grow to several hundreds of thousands of jobs with a

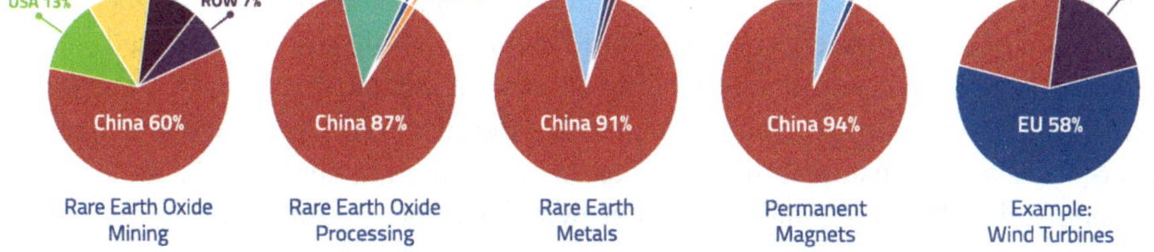

Fig. 7.1 From rare earth extraction to wind turbines. National production shares across the supply chain in 2019 (Gauß et al. 2021)

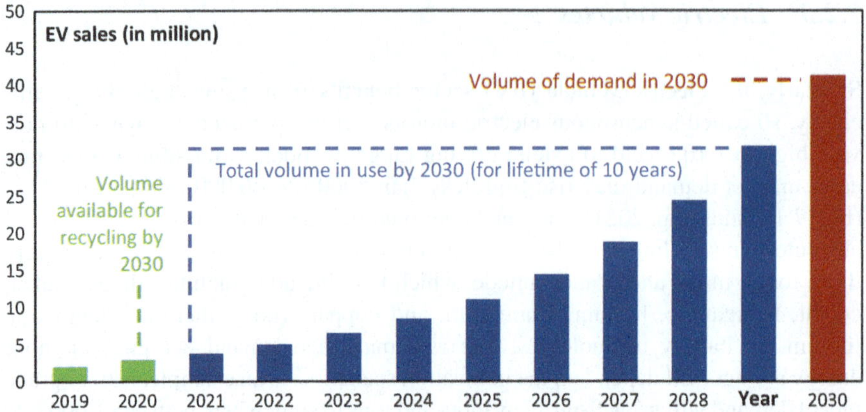

Fig. 7.2 Example calculation of the recycling potential of materials used in electric vehicles (Gauß et al. 2021)

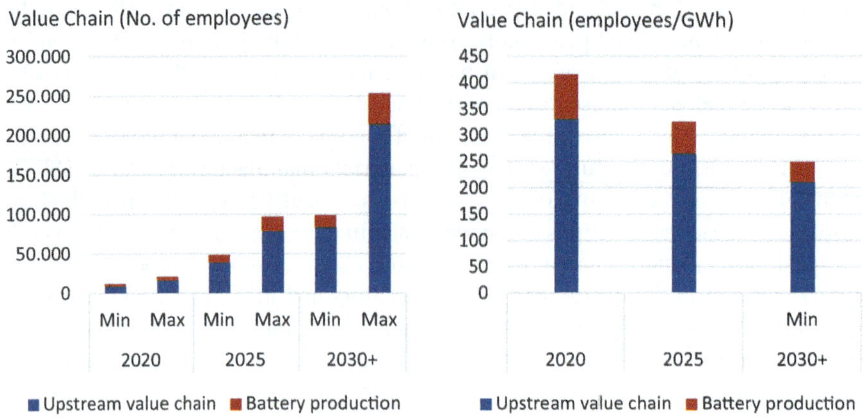

Fig. 7.3 Forecast of potential absolute number of jobs required for the emerging battery industrial sector in Europe, i.e., across the value chain from raw materials extraction and processing ("upstream") to battery production (incl. battery cells and modules). Thielmann et al. 2021

particular demand for jobs upstream, that is, related to materials extraction, processing, and advanced materials manufacturing (Fig. 7.3; Thielmann et al. 2021). If Europe strategically invests in the raw materials value chains required for the implementation of renewable energy and electric mobility it will achieve increased supply chain diversification ultimately leading to cost reduction, increased competition, and more resilient supply chains. In that way, the EU would become a driver of the transparency and sustainability agendas in raw materials and enable the circular economy, which would benefit from a growing materials market to create economic growth and new jobs.

7.3 From Fossil to Bio-Based Resources

A key element of Europe's Green Deal is the transformation from a fossil to a sustainable, circular, and competitive bio-based[1] society (see EU's Green Deal 2019). As a basis, the 2018 EU Bioeconomy Strategy addresses challenges such as meeting the Sustainable Development Goals set by the United Nations and the climate objectives of the Paris Agreement (Kardung et al. 2021). In some countries such as Germany, moreover, a national bioeconomy strategy was pro-actively formulated in 2020 (BMBF, BMEL 2020).

A bioeconomy is circular by nature. CO_2 is recycled from the atmosphere into biomass via photosynthesis. It uses all fractions of renewable resources and keeps the carbon recycled within the life cycle of products. Essential steps towards a sustainable bioeconomy are (after Sillanpää and Ncibi 2017):

- using biomass for bio-based chemicals, polymers, and other material systems
- stimulating bio-based product design with respect to sustainability: durability, reuse, repair, recycling, and biodegradability
- supporting stakeholders along and across new supply chains
- fostering the use of residues and wastes as resource
- intensifying the cooperation with the waste management sector to ensure that the bio-based products can be integrated in collection, separation, recycling, and composting schemes.

The case of Germany can be used to illustrate the challenges of implementing a bioeconomy (Figs. 7.4 and 7.5). Roughly half of Germany's territory is used for agriculture (around 55.7% in 1992, and 46.4% in 2020), approximately one third is forested (29.8% in 1992, and 31.8% in 2020), and an increasing area is covered by settlements, industry, and infrastructure such as roads and rails (11.5% in 1992 and

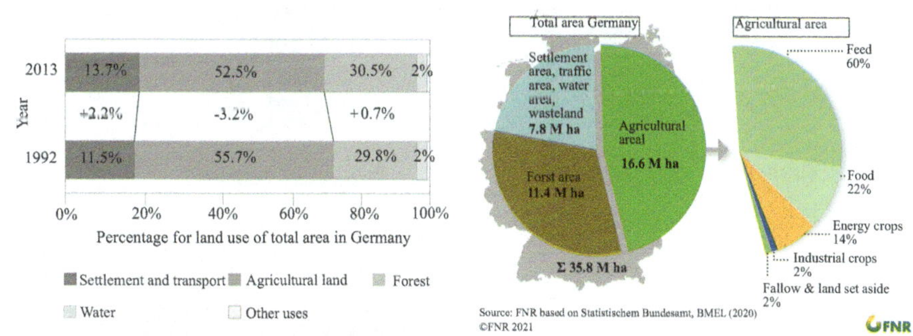

Figs. 7.4 and 7.5 Land use in Germany for the years 1992, 2013, and 2020 (Hoymann et al. 2021; FNR 2022)

[1] Standard EN 16575:2014 defines 'bio-based' as "derived from biomass" and 'bio-based product' as a "product wholly or partly derived from biomass." This standard focuses on areas other than food, feed, and energy applications.

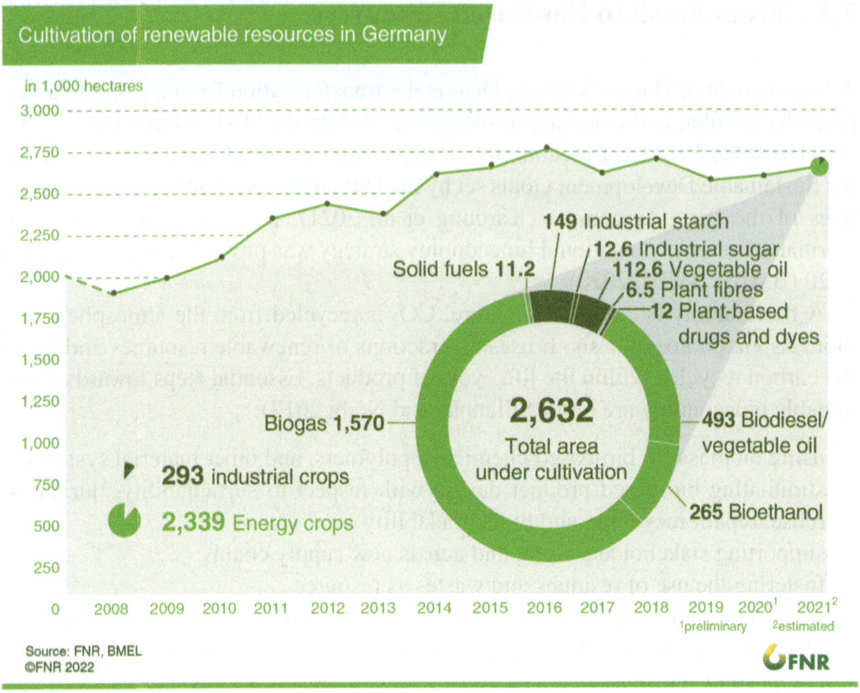

Fig. 7.6 Land use for renewable resources in Germany (FNR 2022)

Fig. 7.7 Types of bio-based products (Stegmann et al. 2020)

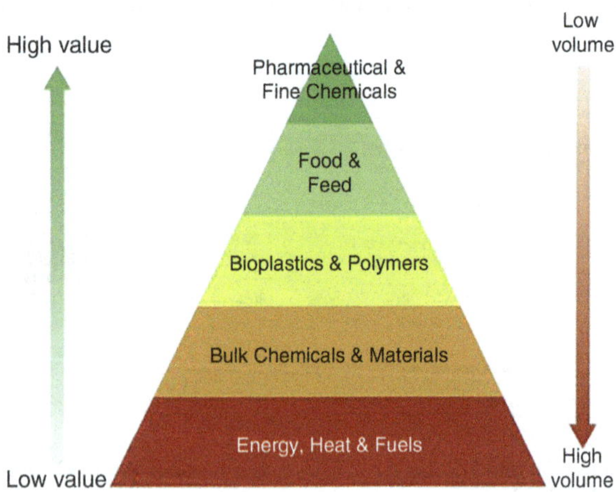

21.8% in 2020). Thus, the land for agriculture has been reduced by more than 9% over the last 30 years.

Of the area actively used for agriculture, more than 80% is used as farmland to grow cereals, food crops, animal fodder, and energy crops such as rapeseed. About 60% of the land is used for livestock farming. The area used for the cultivation of renewable resources in Germany has not changed much during the last 6 years (Fig. 7.6). Today, 90% of renewable resources are used for energetic purposes such as biofuels or biogas and only ca 13% for material use, e.g., starch, oil, and fibers (Brosowski et al. 2016).

Bio-based products can be classified according to a value model, like the waste pyramid of the Circular Economy. This pyramid contains five types of products that differ in value and volume (Fig. 7.7).

The raw materials used by the chemical industry are approximately 50% organic (fossil and bio-based) and 50% inorganic (minerals, metals). Only taking the organic part into account, the overall bio-based share increased from 11% in 2008 to 14% in 2015 (Piotrowski et al. 2016). The 2030 target will be to increase the use of bio-based feedstocks to 25% (Road to Bio 2019).[2] This raises questions and creates challenges and limitations regarding the sustainability of feedstock potential which is discussed in detail in relevant literature (e.g., Hennig et al. 2016).

Research activities are especially needed concerning resource-efficient disruptive technologies for high-value materials on the basis of bio-based feedstocks and

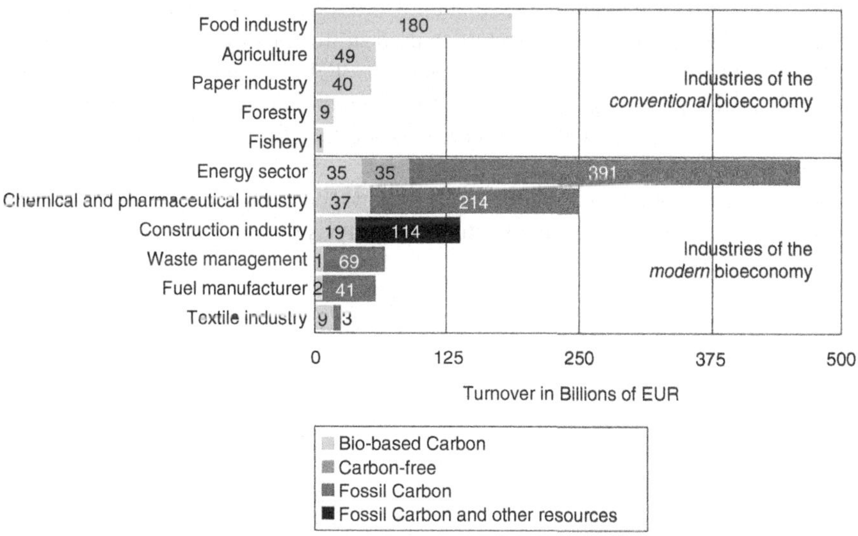

Fig. 7.8 Market shares of different bio-based sectors (in billions of EUR) (Kircher 2020)

[2]Today, R&D activities are supported within the EU, for example, the Bio-Based Industries Joint Undertaking BBI as well as initiatives by individual Member States, such as the initiative "BioEnergie 2021 – Forschung für die Nutzung von Biomasse", *6. Energieforschungsprogramm*, German Hightech Strategy.

natural catalysts. Several examples for successful applications are bio-based lubricants (Syahir et al. 2017), polymers (Babu et al. 2013), and coatings (Noreen et al. 2016). Even high-tech coatings are obtainable with complex property profiles such as functional barrier coatings for food packing (Emmert et al. 2021). An overview on today's markets for bio-based products and their shares is given in Fig. 7.8 (see also Roadmap 2019; Road to Bio 2019). In Germany, industrial fields with growing dynamics are bio-based platform chemicals such as glycerine, sorbitol or succinic acid, bioplastics such as Polyhydroxybutyrat (PHB) and bio-based specialty chemicals, e. g., acrylic acid (Zinke et al. 2016).

The opportunities of using biomass as a renewable resource in industrial production becomes obvious: it has a high potential to substitute fossil-based feedstocks for industry and is circular by nature. Yet, a sustainable and socially acceptable bioeconomy will always have to be balanced out against competing land use, i.e., those linked to agriculture, forestry as well as aspects related to food supply security, climate protection, biodiversity, the use of fertilizers, and the protection of soil and water.

7.4 Towards a Hydrogen Economy

Hydrogen is a secondary energy source and can be produced by electrolysis of acidic or alkaline water solutions. About 4.3–4.9 kWh are required to produce one cubic meter of hydrogen. The electrodes used in this process may contain critical raw materials, like in the case of cathodes made of Ni-Co-Zn or Ni-Mo and anodes made of Ni-La-perovskites or Ni-Co-spinel, which can reduce overvoltage and, thus, improve efficiency. Other technologies are the Proton Exchange Membrane electrolysis (PEM electrolysis) using membranes with Pt as cathode and IrO_2 and Ru-oxihydrate as anode materials. Rare earths doped ZrO_2, CeO_2 or $LaGaO_3$ electrolytes are used in the high temperature electrolysis of water vapour at 800–1000 degree Celsius.

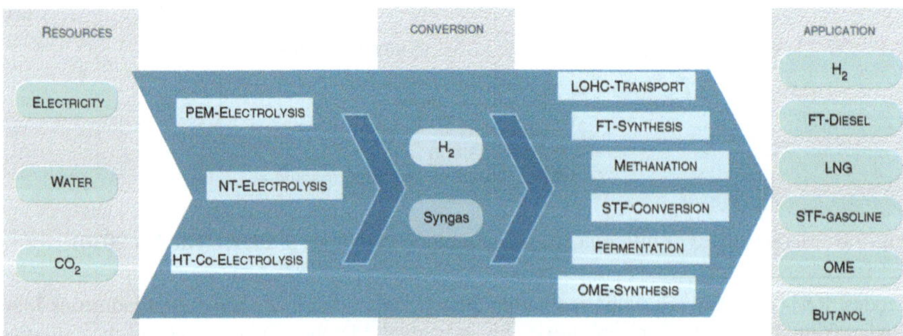

Fig. 7.9 Overview of the PtX-technologies (Ausfelder and Dura 2019)

Table 7.1 Greenhouse Gas (GHG)-Emissions in Germany by sectors (Mio. t CO2-Eqv)

Sector	1990	2015	Target 2030	Target 2050
Energy	427	335	165	21
Industry	284	190	142	14
Traffic	164	165	97	8
Households	132	90	44	7
Agriculture	79	70	54	4
Trade/service	66	38	22	3

Based on Ausfelder and Dura (2019)

Today, approximately 75 million tons of hydrogen are produced worldwide. Only one percent of this hydrogen is made by electrolysis using renewable energy (IRENA 2022). Following the Paris Climate Agreement, several countries have started to promote so-called Power-to-X-technologies (PtX technologies), to increase the *green* production of hydrogen from renewable energy sources like those from wind and solar, and to store oversupplied energy in peak times caused by the intermittent nature of renewable energy. Several innovators consider hydrogen the fundamental molecule of the future, i.e., for the use in the transport sectors as well as in the heavy industry (e.g., so called Green Steel) and the chemical industry sectors (Fig. 7.9; Geres et al. 2020).

Hydrogen and, thus, hydrogen driven technologies are only as green as the energy and the functional materials that are used to run the electrolysis to decompose water. In fact, overall, the generation of energy remains the biggest source of GHG-emissions in Germany (Table 7.1). Yet, the country's energy sector is changing: 2020 was the first year renewable energies represented more than half of the power generated. By 2023, Germany plans to exit from nuclear power, and by 2035 from coal-based power as well.

The second most important producer of GHG-emissions is the German chemical, metal, and manufacturing industry. Iron and steelmaking represent approximately 30% of these emissions. Therefore, various projects already started to substitute coal as the reducing agent. For example, Thyssenkrupp began substituting coke by hydrogen for smelting iron ore in a blast furnace (Thyssenkrupp 2022). A similar project is implemented by SSAB, LKAB, and Vatenfall in Sweden (SSAB 2022). Another project is called WINDH2 in Salzgitter: 30 MW of wind power energy is used for a PEM-electrolysis on-site and direct use for a hydrogen-based reduction process of iron ore, followed by a steel conversion in an electric arc furnace (SALCOS 2022). This process of direct reduction of iron ore (DRI) requires approximately 3.48 MWh/t liquid steel. The aim of all these projects is to trigger a long-term reduction of GHG-emissions in the iron and steel industry by 95% until 2050, presuming renewable energies are sufficiently available and electrical power cost will not exceed 40 €/MWh, which is currently considered to be the upper limit to remain competitive in the international market (Vogl et al. 2018).

Today, hydrogen is already used as fuel in rockets, because of its comparatively high energy density (Fig. 7.10). Next to its potentially low carbon footprint, this is

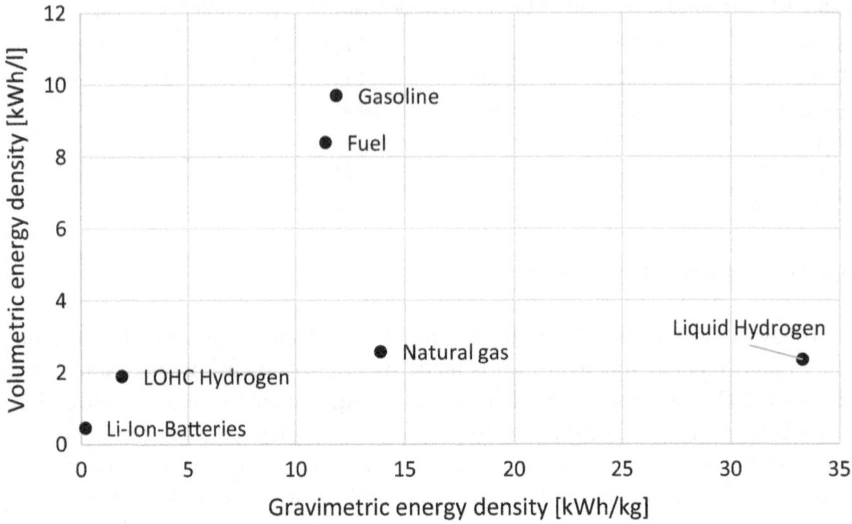

Fig. 7.10 Comparison of energy densities of various media. (Data from Wikipedia and Kuznetsov and Edwards 2010)

one of the key factors why hydrogen is considered as a suitable alternative to fossil fuels to propel vehicles, planes, and ships. Fuel cells are used to convert chemical energy into electricity, which is – often in combination with a battery as an energy storage buffer – used to run an electrical engine. When weight is an important factor, e.g., in aircrafts, hydrogen-based fuel alternatives might become even more important for substituting combustion engines. Indeed, traffic is the third most relevant sector for GHG-emissions in Germany (Table 7.1). While sectors like energy and industry reduced their emissions since 1990 by 22% and 33%, respectively, technological improvements in car engines to decrease CO_2 emissions per car were counterbalanced by an overall increasing number of cars on the streets, and more truck traffic. Still today, 94% of the transport is based on fossil fuels, 4% on biofuel, and only 2% on electrical power (mainly rail). To achieve Germany's environmental goals for the reduction of GHG-emissions, significant efforts in the development of hydrogen and PtX-technologies are necessary to increase the efficiency of the processes and to reduce cost.

As hydrogen is not only interesting for industrial and technical applications but also for the energy and transport sectors, distribution and storage must be studied carefully. Due to its permeability even through metals, losses during storage are significant. Some years ago, the car manufacturer BMW found that its hydrogen pilot car lost a load of hydrogen from its storage tanks within 9 days without usage (Talbot 2006). The key take away was that metal tanks must be lined with special plastics to avoid hydrogen leakage. The research and development community is considering several hydrogen storage alternatives, which include hydrogen storage under pressure and at low temperatures, hydrogen storage by absorption (e.g., metal hydrides), by adsorption (e.g., graphite), and by liquid organic hydrogen carriers (LOHC).

Hydrogen storage under pressure of 700 bar is the more commonly used method in cars until now. The disadvantage is that the compression of the hydrogen gas consumes approximately 12% of its energy. Nevertheless, the Toyota FCHV-car, for example, can reach more than 800 km with one tank volume. Also, storage of liquid hydrogen at temperatures below −240 degree Celsius requires approximately 20% energy consumption; but higher energy densities can be reached. Metal hydrides, such as MgH_2, LiH, Mg(BH4)2, as well as the chemical compound borazane (H_3NBH_3) can be used to store hydrogen. The advantage is a comparably high storage capacity of up to approximately 15% at normal conditions. However, the disadvantage is that of a higher weight. Adsorption of hydrogen requires low temperatures, while studies of suitable materials like carbon nanotubes, TiO_2 nanotubes and SiC nanotubes are ongoing. Due to easier handling and potential use of the existing fuel infrastructure, alternatives called liquid organic hydrogen carriers (LOHC), based on N-ethyl-carbazol or Marlotherm oil have a huge potential for the transport sector: the energy storage capacity could be increased by a factor of five and thus could boost the development of the fuel cell car. This technology seems promising and as a result the first industrial production units are now online in the USA (von Wild et al. 2010).

Recently, Daimler Truck celebrated the test of a liquid hydrogen-based Mercedes-BenzGenH2Truck fuel cells prototype in its plant at Wörth, Germany. In a partnership with Air Liquide, the company tested the filling of two 40 kg tanks with cryogenic liquid hydrogen at −253 degrees Celsius, while material and insulation of the tanks shall provide the higher density of liquid hydrogen over a targeted distance of up to 1000 km. After successful testing, Daimler Truck intends to start production. Also, the first hydrogen powered trains conducted test runs in various parts of Germany. These trains shall cover the future demand on shorter distances without direct supply of electricity.

The most important factor for a successful development of the hydrogen chain will be the competitive availability of renewable electrical power. If, for instance, Germany were to receive the announced 31 TWh of electricity generated by waterpower in Norway, an equivalent of 7.5% of Germany's car fleet could be supplied with green hydrogen. Until 2050, Germany expects an additional power supply for PtX-products in the range of 270 TWh. Hence, not only the production of this electrical power, but also its distribution will require significant investments in the coming 30 years.

Overall, the current consumption of hydrogen in Germany corresponds to about 55 TWh per year. According to the national hydrogen strategy (NWS), Germany expects a ramp up in demand for hydrogen to 90–110 TWh by 2030. Therefore, Germany intends to build several green hydrogen plants (power plants plus electrolysis) of up to 20 TWh power. In addition, decision makers are also considering significant imports of hydrogen, in cooperation with other countries around the North Sea, the Baltics, and in Southern Europe – to make use of wind and solar energy as efficiently as possible.

To achieve its ambitious 2050 targets of a greenhouse gas reduction of 95%, up to 380 TWh additional energy carriers, incl. hydrogen, are required annually. In 2022, Air Liquide announced a project for a 30 MW PEM-electrolyser in Oberhausen based on green electricity. The project is co-financed by the German government and demonstrates the enormous growth and job potential linked to water steam electrolysis.

As indicated above, besides renewable energy, green water electrolysis requires sustainable functional materials containing elements such as magnesium, lithium, boron, cobalt, nickel, gallium, indium, tantalum, and especially platinum group metals and rare earth elements. Most of them are already on the list of the EU Critical Raw Materials (RMIS 2022), because of their high economic importance and the fact that their extraction and processing are highly concentrated in few countries, including China, South-Africa, Turkey, and Indonesia. Considering the enormous investment needs, the responsible sourcing, the sustainable manufacturing, and recycling of respective materials is a must. As an example, the currently forecasted demand for the platinum group metal iridium used in electrolyser anodes would consume 150% of today's respective global primary production in a medium technology deployment scenario modelled by Fraunhofer ISI (Marscheider-Weidemann et al. 2021; Fig. 7.11).

A team of Technical University Munich reported an interesting alternative to platinum-based catalysts (Hardt et al. 2021). They identified an enzyme, which is embedded in a polymer matrix, increases the efficiency of the electrolysis but also the conversion of hydrogen into power. Further R&D-efforts are required to enable higher power densities and thus become competitive to platinum-based technologies.

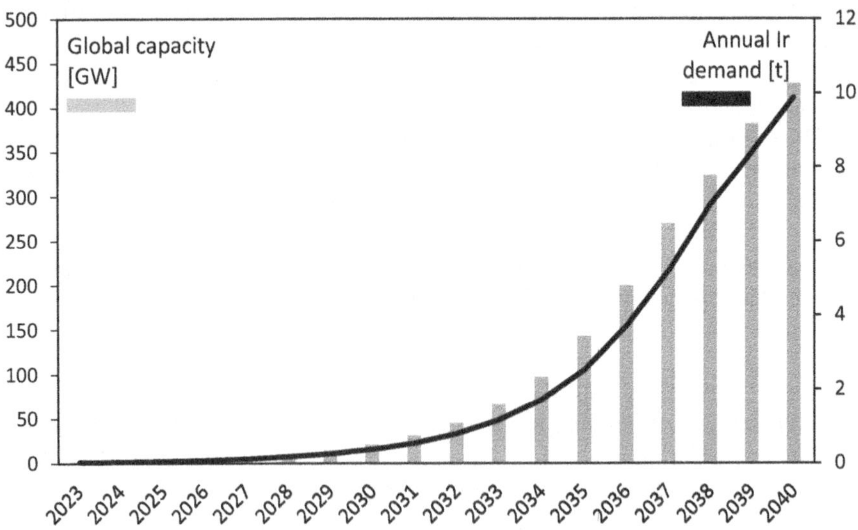

Fig. 7.11 Global electrolyser capacity vs. forecasted annual demand for iridium used in electrolyser anodes. Global Ir production in 2018 was 6.8 t (refined production). After "Medium Scenario (SSP2)" by Marscheider-Weidemann et al. 2021

7.5 A European Response: Innovation and Investments

In its 2021 update, the EU Industrial Strategy underscored that securing greater strategic autonomy and developing more resilient supply chains for Europe's industrial ecosystems are key cornerstones of its policy formulation (JRC 2020). The EU has also implemented a Critical Materials Action Plan which includes, for example, the development of sustainable financing criteria for the mining, extractive, and processing sectors in Delegated Acts on Taxonomy; the mapping of potential supply of secondary critical raw materials from EU stocks and wastes; the identification of mining and processing projects and investment needs and related financing opportunities for critical raw materials in the EU that can be operational by 2025, with a priority focus on promoting new critical raw materials projects in coal-mining regions to facilitate a just transition; and the creation of a dedicated Industrial Alliance on Raw Materials – the European Raw Materials Alliance (ERMA).

The Union also introduced Industrial Alliances as a new policy tool in other sectors. So far, the European Commission has launched a Circular Plastic Alliance, the European Battery Alliance (EBA), the European Clean Hydrogen Alliance, and the Industrial Alliance for Processors and Semiconductor Technologies. Like ERMA, all these Alliances aim to i) identify market gaps and regulatory bottlenecks that hinder innovation breakthrough, as well as ii) promote innovation and investment cases in the sector. The whole process is driven by stakeholder consultation involving industrial players from across a specific value chain as well as representatives from academia and wider society, including NGOs. In the case of the EBA, the most prominent policy measures that were put in place are the EU Strategic Action Plan on Batteries (European Commission 2018) and the update of the EU Battery Directive (European Commission 2022a). In addition, dedicated funding schemes have been set in place. Along with battery related HorizonEurope calls, the *Important Project of Common European Interest* (IPCEI) that constitutes an exceptional policy tool was also launched. The IPCEI has enabled national governments to provide larger amounts of state aid to companies to support the deployment of innovative manufacturing solutions from across the value chain. In the case of the battery sector, two IPCEIs of the size of €3.2 and €2.9 bn of public funding support have been approved.

ERMA, moreover, focuses on promoting competition and innovation to build up the resilience of raw materials and advanced materials supply chains. Two clusters for stakeholder engagement have been defined so far, i.e., on Rare Earth Magnets and Motors (Gauß et al. 2021) as well as Materials for Energy Storage and Conversion (ERMA in print). So far, the ERMA Rare Earth Cluster has identified policy measures that would promote the creation of a level-playing-field in industrial production, the implementation of standards, the promotion of a Circular Economy, and international collaboration. The specific recommendations and actions suggested by the stakeholders were partially reflected in the European Parliament strategy for critical raw materials (EU Parliament 2022), the French raw materials strategy (French Government 2022), the EU Annual Single Market Report

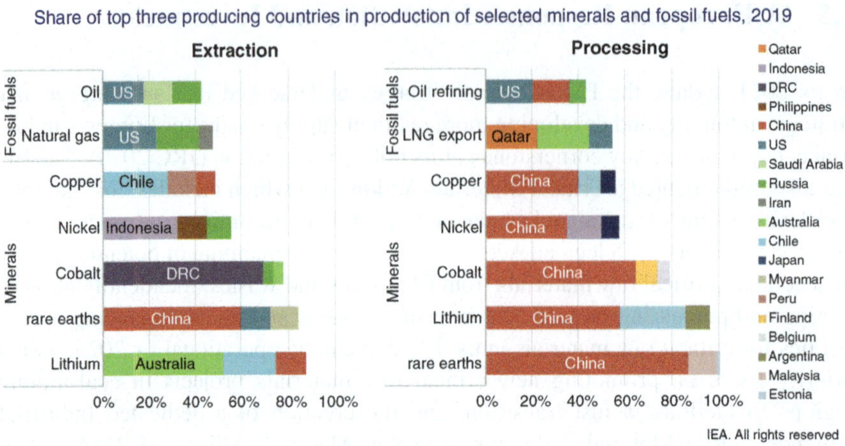

Fig. 7.12 Extraction and processing capacities of key raw materials for the Green Transition (IEA 2021)

2022 (European Commission 2022b), and the Critical Raw Materials Act 2022 (European Commission 2022c). Strategic investment cases have also been identified, related to primary and secondary raw materials projects, which would have the potential to partially re-shore this vital industry to Europe (Fig. 7.12).

7.6 Conclusions

The global green transition will continue to accelerate but as the chapter has demonstrated, it produces numerous disruptive changes in a wide range of industrial ecosystems. The European Union, among the most committed economies to the decarbonization process, has acknowledged the challenges and risks. Importantly, it has also responded decisively by creating industrial alliances on hydrogen, batteries, semiconductors, and raw materials to address regulatory bottlenecks and spotlight investment opportunities. Still, there is much more work that needs to be done for a true green transition to materialize both inside of the EU but also in partner nations throughout the developing world. In the end, to succeed, the green transition must be just, but also make economic sense. Throughout the chapter, the sector-by-sector analysis has indicated that the green transition will be more costly than countries including the EU have estimated and have been willing to openly acknowledge. Renewable energy and electric mobility are only possible by securing substantial amounts of sustainable raw materials and advanced materials at low-cost and along resilient, hence diversified and circular, supply chains. The ramp up of a bioeconomy that, for instance, envisions the large-scale replacement of traditional fossil-based feedstocks to the chemical industry by zero-emission and circular biomaterials needs to consider the boundaries of competing land use. A hydrogen economy is only as green as the electricity provided to produce hydrogen. Its sustainable scalability also fundamentally depends on countries' capacities to build technologies and infrastructures at large scale using materials with low environmental and social footprints. Therefore, moving forward, the European Union should consider the following key aspects:

1. Create a level-playing-field to promote EU industrial competitiveness on a global level, which is a key driver for innovation: Stakeholder buy-in will become more important than ever before. Policy makers will not be able to make decisions in a vacuum and behind closed doors, even though the urgency of the climate crisis dictates accelerated action. The socio-economic implications and environmental footprint of this transition are such that they require new standards and regulatory frameworks if they are to align with EU declarations for a sustainable and just economic transformation. Governments and industry will need to continue to strategically, but also generously, invest in sustainable raw material value chains that will enable a successful transition. This is even more important for the production of biomass, hydrogen, and strategic raw materials, for which new agreed up frameworks will be essential and should include a careful and honest assessment of impacts related to competitiveness (material cost and performance) as well as sustainability (environment, climate, people, and governance). These frameworks will need to ensure the competitiveness of the European market to avoid the migration of significant parts of the industry to other parts of the world, which may not follow a similar sustainability agenda (consider, e.g., incentives, fiscal measures, carbon border tax mechanism, match financing using public grants and loans to trigger large private investments).
2. Invest in renewable energy: Massive amounts of renewable energy will be required to produce green hydrogen but also the raw materials and advanced materials necessary to run the full value chain of clean technology materials, i.e., from exploration, mining, and processing to product engineering and recycling.
3. Promote a Circular Economy: Critical and strategic materials are indispensable for the transition. Like the rest of the world, Europe needs to invest in the full value chain of raw materials, including exploration and mining, to diversify supply, share the environmental burden, and be the leader in responsible sourcing as well as sustainable materials manufacturing and the circular economy.
4. Boost Open R&D&I communities: A vivid research and innovation community has been a competitive advantage for European industry and it should continue to be strengthened. Co-financed public private partnerships are an ideal tool to link research to industrial needs.
5. Identify and promote strategic partnerships: It is essential to secure the supply of sustainable raw materials, hydrogen, and electricity via international partnerships. In addition, international cooperation on R&D&I and technology transfer via research institutions and companies are key.

Though the risks of this unprecedented industrial and infrastructure shift may, at the moment, seem exceptionally high, the EU has already become one of the most active drivers of the circular economy to power the green transition and the hope is that staying the course will further benefit the economy, create well-paying jobs, and help the EU achieve its goal of a just climate neutrality by 2050.

References

Agora Verkehrswende. 2019. Klimabilanz von Elektroautos. Einflussfaktoren und Verbesserungspotenzial. https://www.agora-verkehrswende.de/veroeffentlichungen/klimabilanz-von-elektroautos/. Accessed 26 November 2022.

Ausfelder, F., and H. Dura. 2019. Optionen für ein nachhaltiges Energie-System mit Power-to-X-Technologien, DECHEMA. https://dechema.de/dechema_media/Downloads/Positionspapiere/DEC_P2X_II_V06_Online_small.pdf. Accessed 25 November 2022.

Babu, R.P., K. O'Connor, and R. Seeram. 2013. Current Progress on bio-based polymers and their future trends. *Progress in Biomaterials* 2: 8. https://link.springer.com/article/10.1186/2194-0517-2-8. Accessed 27 November 2022.

Brosowski, A., D. Thraen, U. Mantau, B. Mahrod, G. Erdmann, P. Adler, W. Stinner, G. Reinhold, T. Hering, and C. Blank. 2016. A review of biomass potential and current utilisation – Status quo for 93 biogenic wastes and residues in Germany. *Biomass and Bioenergy* 95: 257–272. https://doi.org/10.1016/j.biombioe.2016.10.017. Accessed 20 November 2022.

Bundesministerium für Bildung und Forschung BMBF, Bundesministerium für Ernährung und Landwirtschaft BMEL, eds. 03–2020. Nationale Bioökonomiestrategie der Bundesregierung, https://biooekonomie.de/themen/politikstrategie-deutschland. Accessed 23 November 2022.

Council of the European Union. 2019. Preparations for the United Nations Framework Convention on Climate Change (UNFCCC) meetings (Santiago de Chile, 2–13 December 2019) – Council conclusions: https://data.consilium.europa.eu/doc/document/ST-12796-2019-REV-1/en/pdf. Accessed 24 November 2022.

DERA. 2021. Nickel. https://www.deutsche-rohstoffagentur.de/DERA/DE/Aktuelles/rohstoff_Nickel.html. Accessed 31 October 2022.

Emmert, K., S. Amberg-Schwab, F. Braca, A. Bazzichi, A. Cecchi, and F. Somorowsky. 2021. bioORMOCER® — Compostable functional barrier coatings for food packaging. *Polymers* 13: 1257. https://doi.org/10.3390/polym13081257. Accessed 24 November 2022.

ERMA. in print. *Materials for energy storage and conversion*. A European Call for Action.

European Commission. 2018. Sustainable Mobility for Europe: Safe, connected and clean. ANNEX 2 – Strategic Action Plan on Batteries. https://eur-lex.europa.eu/resource.html?uri=cellar:0e8b694e-59b5-11e8-ab41-01aa75ed71a1.0003.02/DOC_3&format=PDF. Accessed 12 November 2022.

———. 2019. A European Green Deal – Factsheets on the European Green Deal. https://commission.europa.eu/publications/factsheets-european-green-deal_en

———. 2020. Study on the EU's list of Critical Raw Materials (2020). Critical Raw Materials Factsheets (Final). https://rmis.jrc.ec.europa.eu/uploads/CRM_2020_Factsheets_critical_Final.pdf. Accessed 12 October 2022.

———. 2021. Delivering the European green deal. https://ec.europa.eu/info/strategy/priorities-2019-2024/european-green-deal/delivering-european-green-deal_en. Accessed 22 October 2022.

———. 2022a. Commission publishes evaluation of the EU Batteries Directive. https://ec.europa.eu/info/news/commission-publishes-evaluation-eu-batteries-directive-2019-apr-09-0_en. Accessed 22 October 2022.

———. 2022b. Annual single market report. https://ec.europa.eu/docsroom/documents/48877. Accessed 13 October 2022.

———. 2022c. The Critical Raw Materials Act. https://ec.europa.eu/commission/presscorner/detail/en/STATEMENT_22_5523. Accessed 11 October 2022.

European Parliament. 2022. European Parliament resolution of 24 November 2021 on a European strategy for critical raw materials (2021/2011(INI)).

Fachagentur Nachwachsende Rohstoffe FNR. 2022. Cultivation of renewable resources in Germany. https://statistik.fnr.de/download.php. Accessed 22 October 2022.

French Government. 2022. Investir dans la France de 2030: remise au gouvernement du rapport Varin sur la sécurisation de l'approvisionnement en matières premières minérales et ouverture d'un appel à projets dédié. https://www.gouvernement.fr/investir-dans-la-france-de-2030-remise-au-gouvernement-du-rapport-varin-sur-la-securisation-de-l#:~:text=Barbara%20Pompili%2C%20ministre%20de%20la,septembre%202021%20%C3%A0%20Philippe%20Varin. Accessed 29 November 2022.

Gauß, R., et al. 2021. Rare earth magnets and motors: A European call for action A report by the rare earth magnets and motors cluster of the European Raw Materials Alliance. https://erma.eu/app/uploads/2021/09/01227816.pdf. Accessed 13 November 2022.

Geres, R., A. Kohn, S. Lenz, F. Ausfelder, A.M. Bazzanella, and A. Möller. 2020. Roadmap Chemie 2050, Auf dem Weg zu einer treibhausgasneutralen chemischen Industrie in Deutschland. DECHEMA. https://www.vci.de/vci/downloads-vci/publikation/2019-10-09-studie-roadmap-chemie-2050-treibhausgasneutralitaet.pdf. Accessed 11 November 2022.

Gregoir, L. 2022. Metals for clean energy: Pathways to solving Europe's raw materials challenge. Eurometaux. https://eurometaux.eu/media/jmxf2qm0/metals-for-clean-energy.pdf. Accessed 17 November 2022.

Hardt, S., S. Stapf, D.T. Filmon, J.A. Birrell, O. Rüdiger, V. Fourmond, Ch. Léger, and N. Plumeré. 2021. Reversible H_2 oxidation and evolution by hydrogenase embedded in a redox polymer film. *Nature Catalysis* 4: 251–258. https://doi.org/10.1038/s41929-021-00586-1. Accessed 20 October 2022.

Hennig, C., A. Brosowski, and S. Majer. 2016. Sustainable feedstock potential – A limitation for the bio-based economy? *Journal of Cleaner Production* 123: 200–202. https://doi.org/10.1016/j.jclepro.2015.06.130. Accessed 14 November 2022.

Hoymann, J., S. Baum, P. Elsasser, R. Dechow, M. Gutsch, and J. Fick. 2021. Ist-Situation der Landnutzung in Deutschland. In *Wechselwirkungen zwischen Landnutzung und Klimawandel*, ed. H. Goemann and J. Fick. Wiesbaden: Springer Spektrum. https://doi.org/10.1007/978-3-658-18671-5_2. Accessed 14 November 2022.

IEA. 2021. The role of critical minerals in clean energy transitions. Part of World Energy Outlook. https://iea.blob.core.windows.net/assets/ffd2a83b-8c30-4e9d-980a-52b6d9a86fdc/TheRoleofCriticalMineralsinCleanEnergyTransitions.pdf. Accessed 01 November 2022.

Intergovernmental Panel on Climate Change. 2021. Climate change widespread, rapid, and intensifying – IPCCC. https://www.ipcc.ch/2021/08/09/ar6-wg1-20210809-pr/. Accessed 12 November 2022.

JRC. 2020. The role of rare earth elements in wind energy and electric mobility. https://publications.jrc.ec.europa.eu/repository/handle/JRC122671. Accessed 03 November 2022.

IRENA. 2022. Hydrogen. https://www.irena.org/Energy-Transition/Technology/Hydrogen

Kardung, M., K. Cingiz, O. Costenoble, R. Delahaye, W. Heijman, M. Lovric, M. van Leeuwen, R. M'Barek, H. van Meijl, S. Piotrowski, T. Ronzon, J. Sauer, D. Verhoog, P.J. Verkerk, M. Vrachioli, J.H.H. Wesseler, and B.X. Zhu. 2021. Development of the circular bioeconomy. Drivers and Indicators. *Sustainability* 13: 413. https://doi.org/10.3390/su13010413.

Kircher, M. 2020. Der Stand der Bioökonomie in Deutschland (pp. 39–60). In: Weg vom Öl. Springer, Berlin, Heidelberg. https://doi.org/10.1007/978-3-662-61490-7_4. Accessed 14 October 2022.

Kuznetsov, V.L., and P.P. Edwards. 2010. Functional materials for sustainable energy technologies: Four case studies. *ChemSusChem* 2010 (3): 44–58. https://doi.org/10.1002/cssc.200900190. Accessed 24 November 2022.

Marscheider-Weidemann, F., S. Langkau, E. Eberling, L. Erdmann, M. Haendel, M. Krail, A. Loibl, Ch. Neef, M. Neuwirth, L. Rostek, S. Shirinzadeh, D. Stijepic, L. Tercero Espinoza. 2021. Rohstoffe für Zukunftstechnologien 2021. DERA Rohstoffinformationen 50.

Noreen, A., K. Mahmood Zia, M. Zuber, S. Tabasum, and A. Fawad Zahoor. 2016. Bio-based polyurethane: An efficient and environment friendly coating systems: A review. *Progress in Organic Coatings* 91: 25–32. https://doi.org/10.1016/j.porgcoat.2015.11.018. Accessed 11 October 2022.

Piotrowski, S., M. Carus, and D. Carrez. 2016. European bioeconomy in figures. *Industrial Biotechnology* 12 (2): 78–82. https://doi.org/10.1089/ind.2016.29030.spi and updated 2019: https://biconsortium.eu/downloads/bioeconomy-figures-2008-2016-update-2019. Accessed 08 November 2022.

RMIS. 2022. CRM list 2020. https://rmis.jrc.ec.europa.eu/?page=crm-list-2020-e294f6. Accessed 21 September 2022.

Road to Bio. 2019. Roadmap for the Chemical Industry in Europe towards a Bioeconomy. https://roadtobio.eu/uploads/publications/roadmap/RoadToBio_strategy_document.pdf. Accessed 21 September 2022. Accessed 24 October 2022.

SALCOS. 2022. Das Konzept für eine nachhaltige Zukunft. https://salcos.salzgitter-ag.com/de/. Accessed 24 November 2022.

Sillanpää, M., and C. Ncibi. 2017. *A sustainable bioeconomy – The green industrial revolution.* Springer; ISBN: 978-3-319-55635-2.

SSAB. 2022. HYBRIT. Eine neue, bahnbrechende Stahlerzeugungstechnologie. https://www.ssab.com/de-de/fossil-free-steel/hybrit-a-new-revolutionary-steelmaking-technology

Stegmann, P., M. Londo, and M. Junginger. 2020. The circular bioeconomy: Its elements and role in European bioeconomy clusters. *Resources, Conservation & Recycling: X* 6: 100029. https://doi.org/10.1016/j.rcrx.2019.100029. Accessed 19 October 2022.

Syahir, A.Z., N.W.M. Zulkifli, H.H. Masjuki, M.A. Kalam, A. Alabdulkarem, M. Gulzar, L.S. Khuong, and M.H. Harith. 2017. A review on bio-based lubricants and their applications. *Journal of Cleaner Production* 168: 997–1016. https://doi.org/10.1016/j.jclepro.2017.09.106. Accessed 23 November 2022.

Talbot, D. 2006. Unterwegs im Wasserstoff-7er, MIT technology review. https://www.heise.de/hintergrund/Unterwegs-im-Wasserstoff-7er-279187.html. Accessed 18 November 2022.

Thielmann, A., Ch. Neef, T. Hettesheimer, K. Ahlbrecht, and S. Ebert. 2021. *Future expert needs in the battery sector*. EIT RawMaterials. https://eitrawmaterials.eu/wp-content/uploads/2021/03/EIT-RawMaterials-Fraunhofer-Report-Battery-Expert-Needs-March-2021.pdf

Thyssenkrupp. 2022. Injection of hydrogen into blast furnace: thyssenkrupp Steel concludes first test phase successfully. https://www.thyssenkrupp-steel.com/en/newsroom/press-releases/thyssenkrupp-steel-concludes-first-test-phase-successfully.html. Accessed 10 November 2022.

United Nations. 2015. The Paris Agreement on climate change. https://unfccc.int/process-and-meetings/the-paris-agreement/the-paris-agreement. Accessed 27 October 2022.

von Wild, J., T. Friedrich, A. Cooper, B. Toseland, G. Muraro, W. TeGrotenhuis, Y. Wang, P. Humble, and A. Karim. 2010. Liquid organic hydrogen carriers (LOHC): An auspicious alternative to conventional hydrogen storage technologies. *18th World Hydrtogen energy conference 2010 – WHEC 2010. Proceedings of he WHEC*, eds. D. Stolten, Th. Grube, May 16–21. 2010, Essen. https://juser.fz-juelich.de/record/135562/files/HS2b_2_von-Wild.pdf. Accessed 03 November 2022.

Vogl, V., Ahman, and M. Nilsson, L.J. 2018. Assessment of hydrogen direct reduction for fossil-free steelmaking. *Journal of Cleaner Production* 203: 736–745. https://doi.org/10.1016/j.jclepro.2018.08.279. Accessed 12 February 2023.

WindEurope. 2022. Why is the European wind industry struggling? https://windeurope.org/intelligence-platform/product/why-is-the-european-wind-industry-struggling/. Accessed 18 November 2022.

Wirtschaftswoche. 2019, November 29. Altmaier erwartet 2024 serienmäßige Batteriezellen-Produktion in Deutschland. https://www.wiwo.de/politik/deutschland/elektromobilitaet-altmaier-erwartet-2024-serienmaessige-batteriezellen-produktion-in-deutschland/25284062.html. Accessed 28 October 2022.

Zinke, H., B. El-Chichakli, P. Dieckhoff, S. Wydra, and B. Hüsing. 2016. Bioökonomierat: Bioökonomie für die Industrienation – Ausgangslage für biobasierte Innovationen in Deutschland verbessern. https://www.biooekonomie-bw.de/download_file/force/15984/75215. Accessed 21 September 2022.

Roland Gauß is head of Innovation and Business Intelligence at EIT RawMaterials. He received his PhD from the University of Tübingen in 2008. His research interests are related to metallurgy, the resource criticality and life cycles of materials, and innovation. He worked in the fields of Nd-Fe-B processing and recycling, the extractive metallurgy of copper, and geochemical finger-printing. He was Marie Curie Research Fellow at University College London and Head of Department at Fraunhofer IWKS, and joined EIT RawMaterials as Senior Advisor in 2016. He coordinates the Cluster Rare Earth Magnets and Motors of the European Raw Materials Alliance.

Carsten Gellermann is a chemist and received his PhD from the Julius-Maximilians-Universität of Würzburg, Germany. From 1994, he worked at the Fraunhofer Institute for Silicate Research ISC in Würzburg, initially as a project manager in the field of materials development, and from 2007, he was head of the "Particle Technology and Interfaces" working group. After joining the Fraunhofer Project Group Materials Recycling and Resource Strategy IWKS in Alzenau and Hanau, he took over as head of the business unit "Slags, Sludges, Landfills" in 2012. From 2016 onward, he was responsible for the "Secondary Raw Materials" business division in Alzenau, focusing on business and technology developments, e.g., for the recycling of strategic resources and nutrients. In the meantime, he has returned at the Fraunhofer Institute for Silicate Research in Würzburg and is, among other things, managing director of the Fraunhofer Network Nanotechnology.

Alexander Maurer received his PhD in the Sciences from the University of Heidelberg. He has years of experience in the global business of industrial, food, and pharmaceutical additives and fine chemicals, their technologies, and production processes as well as in the competitive landscape and the sourcing of raw materials. He works as an independent consultant on recycling projects, future technologies, and process developments. He is a member of the supervisory board of BK Giulini GmbH and a number of industrial and scientific associations.

Part III
Mining and Environmental Impacts

Chapter 8
Social and Environmental Impacts of Rare Earth Mining, Processing, and Proliferation: Scoping and Preliminary Assessment

Julie Michelle Klinger

Abstract This chapter uses qualitative, historical, environmental, and epidemiological research from a range of sectors to outline the basis of comparison between *ex ante* and *ex post* pollution controls in the rare earth sector. Following an introduction of the problem, this chapter reviews the contemporary social and environmental impacts of rare earth mining, mainly in China, and considers the political economic frameworks that can limit effective mitigation of social and environmental harm as well as supply chain diversification. It then assesses and provides a preliminary scope for evaluating the social and environmental costs of mining within and beyond the site of extraction, and on this basis compares the possible cost parameters for *ex post* vs. *ex ante* mitigation of social and environmental harms. A brief discussion of limitations and areas of critical research need concludes the chapter.

Keywords Baotou · Environmental regulation · Rare earths · Contamination · Mining · Green nationalism

8.1 Introduction

The first major news stories of the environmental and social devastation caused by rare earth mining in China circulated in Anglophone media in 2009 (Margonelli 2009). Images of massive tailings ponds into which pipes belched mineral and chemical slurries against an industrial skyline in northern China were published alongside photos of peasant miners knee-deep in the muck of informal and illegal mines in the southeast. These stories generally either condemned China for environmental irresponsibility or guilted western tech consumers and renewable energy

J. M. Klinger (✉)
Geography and Spatial Sciences, University of Delaware, Newark, DE, USA
e-mail: klinger@udel.edu

S. Kalantzakos (ed.), *Critical Minerals, the Climate Crisis and the Tech Imperium*, Archimedes 65, https://doi.org/10.1007/978-3-031-25577-9_8

advocates whose demand for rare earth-bearing technologies was blamed for fueling such devastation (Butler 2012; Fisher and Fitzsimmons 2013; Jones 2010), ignoring that the primary application for rare earths in the US is in chemistry applications and in petroleum refining.

These stories, combined with a sudden price increase in the latter months of 2010, sparked a brief period of 'green nationalism,' in which interest groups on opposing sides of the environmentalist, mining, and industrial policy debates in the West called for reinvestment in rare earth supply chains outside of China (Klinger 2017, pp. 142–146), with limited results. In the meantime, central government initiatives in China were already underway to transform much of the domestic rare earth industry by closing informal and small-scale mines, consolidating private companies into state-owned enterprises, and by shifting from an emphasis on mining to value added processing. In 2018, China became a net importer of rare earth elements for the first time since 1985. Through a host of programs, the central government has invested considerable resources in clean-up, public health programs, and environmental remediation. Although the precise budget for this is difficult to confirm, the impacts of such activities—visible in deactivated mining sites, removal of contaminated soil, and the construction of specialized medical facilities—attest to a significant effort.

Outside of China, initiatives to expand rare earth production have encountered a variety of obstacles. Most fundamentally, aside from the brief period of green nationalism in the early 2010s, most downstream firms have been content to source from the lowest-price producer, which continues to be China. Many proposals to garner government subsidies to support new initiatives outside of China have had uneven results. Often, environmental regulation is blamed for this state of affairs: regulation increases the cost of production while forcing firms to compete in a cost-competitive environment. This begs the question as to whether an up-front investment in mitigating social and environmental hazards that may result from rare earth mining and processing is more or less expensive than cleaning up after the fact. This deceptively straightforward question raises a host of other questions, particularly around how costs, and responsibilities for them, are defined and distributed.

Up-front investments in mitigating the social and environmental harms of mining continue to be controversial. On one hand they are demanded by frontline communities, environmental and public health experts, laborers, consumers, and increasingly by downstream manufacturers responding to political pressure to clean up, or "green" their supply chains. On the other hand, they continue to be actively resisted by some members of industry, their allies in scientific communities, and outright discouraged by political and economic frameworks that prioritize downward price competition above all else. The argument against comprehensive up-front, or *ex ante,* investments in mitigating social and environmental harms tends to revolve around a stylized cost competitiveness analysis that often obscures the costs borne by the public before, during, and after the life of the mine. This is because the environmental and public health effects of large-scale mining operations, once they occur, are so temporally and spatially expansive that full clean-up is extremely difficult, if not impossible. Impacts occurring further along the supply chain are seldom considered. Instead, remediation criteria are determined, often after a lengthy struggle, according to 'optimal' clean-up strategies (Lappi 2018). What constitutes 'optimal' is generally defined according to economic rather than environmental or public health priorities.

It is possible that *ex ante* investments to limit the dispersal of pollutants in the first place are far less expensive, even if they are often blamed for overruns and extensions that result in production delays for individual firms. This is because they nevertheless occur within a limited spatial and temporal horizon of the mining and processing sites as compared with the decades-long investments involved in an *ex post* clean-up of contamination resulting from long-term or perhaps poorly controlled operations. This rich debate surrounding the question not of what is ethical or environmentally responsible, but simply which is cheaper—*ex ante* or *ex post* pollution controls to mitigate social and environmental harms from mining and processing activities—has yet to be subjected to a comprehensive quantitative analysis.

The initial motivation for this chapter was precisely that: to conduct a straightforward quantitative analysis comparing the cost of cleaning up current and former mining sites with the cost of mandating and implementing social and environmental protections from the start. Anticipating diverse and scattered data sources, I planned to draw on examples from around the world, and to extrapolate costs tallied for other extractive operations—such as gold, uranium, copper, coal, or arsenic—to develop some estimates for the rare earth sector. The purpose was to assess whether implementing social and environmental safeguards in rare earth mining and processing at the outset is more or less expensive than business as usual, which involves *ex post* clean-up, or, depending on the context, no remedial action at all.

The prevailing political economy limits the scope to which industries are able or required to control pollution, whether through limited regulation or limited support for safeguards in a competitive market economy. This situation therefore has effectively outsourced the cost of clean-up and maintenance to the public after problems have accumulated, and often well after the life of a given mining company. Regulatory controls, remediation plans, and restoration technologies continue to be insufficient to address the problem (Pepper et al. 2021; Shah and Davery 2020; Sun et al. 2018; Unger et al. 2020). Even in the best cases, a long-term public burden is nearly always imposed by extractive activities whether in the form of direct clean-up of disasters (Owen et al. 2020), or in the despoilation of lands that performed other valuable ecosystem services, such as carbon sequestration, erosion control, or the support of agrarian or other sustainable livelihoods (Lawer et al. 2019; Zhang et al. 2021). A further consideration is where the waste materials from mine site remediation are relocated to—in short, even the most thorough clean-up activities involve the creation of other waste sites, which generates their own direct and indirect monetary costs that are most often absorbed by the public.

In attempting to assemble and tally these costs, I instead found a sparse data landscape populated by government reports, the occasional news story, technical case studies, and remediation research findings (e.g. EPA 1997; Liebenberg et al. 2012; Naidu et al. 2019). Although rich in many other kinds of information, very little cost data was disclosed. This means that it is not currently possible to conduct a straightforward quantitative analysis. Two conclusions can be drawn from this. First, that arguments against *ex ante* controls are not based on quantitative data, and second, that the actual cost of addressing the impacts of mining, processing, and distributing rare earths—or other mineral commodities for that matter—remains elusive. However, a qualitative and multi-method analysis enables a comparative scoping assessment of the different ranges of costs involved in *ex ante* and *ex post* pollution controls. This provides a pathway for further analysis, even if dollar amounts cannot be neatly compared as of this writing.

Three primary reasons explain why, despite the diversity and extent of remediation at mining sites throughout the world, the costs are not well-documented. First, clean-up may be undertaken in response to compliance mandates and enforcement actions, disasters, or local permitting requirements. In the US, while specific actions taken must be reported, disclosure of cost data usually is not required. Second, actions vary in spatial and temporal scope, the limits of which are subject to contestation between firms, regulators, and frontline communities, each of whom works with differing notions of risk and cost. Clean-up actions range from one-off responses, like cleaning up a minor spill, to emergency operations following a major disaster, to complex and long-term programs that are completed in several phases over years or decades. Costs may be borne by multiple entities, and cumulative cost data is seldom readily available. Third, costs are often considered proprietary to the mine operator (EPA 1997). Because precise calculations are inhibited by the fact that cost information is fragmented or held as confidential, public capacity to assess various options and make informed decisions is undermined. This casts doubt over the effective capacity of policy-makers, analysts, or investors to undertake accurate cost-benefit analyses of different pollution scenarios.

Nevertheless, from the scattered cost data that is available (but, it must be noted, difficult to verify), it is possible to surmise that even in the context of enforcement operations, or in response to disasters, the spatial and temporal extent of social and environmental harms are defined in a limited rather than comprehensive manner, determined more according to the demarcated extent of jurisdictions and liabilities rather than the actual extent and full effects of pollutant dispersal. This spatial and temporal circumscription is in the interest of both regulatory and industry actors: regulatory actors require clear jurisdictional limits for more effective project formulation, implementation, and budget performance. Industry actors seek to impose spatial and temporal limits in order to reduce liabilities (Wong et al. 1999). Therefore, even if the dollars and cents required for mine site reclamation, remediation, and clean-up were fully disclosed and easily accessible to researchers, the data would most likely not represent a full picture of the social and environmental impacts of mining and processing (e.g. Beckett and Keeling 2018).

There is no consensus on what would constitute a 'full picture' of the social and environmental impacts of mining and processing. Hence the discussion in this chapter explores this question instead, drawing on qualitative, historical, environmental, and epidemiological research from a range of mineral sectors and public health concerns to outline a potential basis of comparison between *ex ante* and *ex post* pollution controls. The broad implications are that up-front investments in mitigating social and environmental harms are far less expensive, even if they subject individual firms to overruns and extensions, because they nevertheless occur within a limited time horizon compared with the decades-long investments involved in the clean-up of contaminated sites and the multi-generational health impacts that can result from exposure. Even in the most intensive remediation and reclamation cases documented, full landscape restoration is nearly impossible, and most severe health effects are irreversible. Local inhabitants, laborers, expropriated peoples, and downstream communities can continue to bear profound costs for generations, which no amount of money could reverse. But simply because profound impacts elude full remediation via monetary compensation does not mean that their cost is equal to zero. Profound impacts generate other costs, some of which are considered below as part of the scoping exercise.

The remainder of the chapter provides background to contemporary social and environmental impacts of rare earth mining, mainly in China, the dominant Anglophone characterization of which limits policy discussion on rebuilding a rare earth supply chain in North America or Western Europe. It then assesses and provides a preliminary scope for evaluating the social and environmental costs of mining within and beyond the site of extraction, and on this basis compares the possible cost parameters for *ex post* vs. *ex ante* mitigation of social and environmental harms. A brief discussion of limitations and areas of critical research need is presented in the conclusion.

8.2 Background: Tracing the Environmental Footprints of the Rare Earth Industry

For more than a decade, lurid photos of environmental devastation in China have circulated globally to emphasize and often moralize on the costs of the country's impressive economic growth. In the rare earth sector, these images served a variety of purposes. A limited and specific set of photos taken from a tailings pond located to the southwest of Baotou municipality in Inner Mongolia Autonomous Region was widely republished with attention-grabbing headlines about "toxic lakes" and "green energy's dirty little secret." In the immediate aftermath of the 2010 rare earth crisis, this fueled a kind of "Green Nationalism" in the US, whereby "beating China" in the rare earths game would be achieved through environmentally superior practices (Klinger 2017). Hundreds of ideas were floated, millions of dollars raised, and only a very small handful of initiatives managed to go beyond the drawing board (Cox and Kynicky 2018).

A measure of alarm over the social and environmental impacts of rare earth mining and processing that had accumulated in some regions of China was certainly warranted. As detailed in my book (Klinger 2017), pollution from over 40 years of rare earth mining and processing created landscapes of environmental and epidemiological devastation, caused by the contamination of soil and water, which despoiled agricultural lands, sickened animals, and led to a widespread phenomenon of neurological, orthopedic, cognitive, reproductive, respiratory, and degenerative illnesses among the people of Baotou and vicinity. This added to the fatalities resulting from the occupational hazards of working in mining and processing facilities. Although these effects had been building for many years, they did not capture the attention of general Anglophone audiences until the rare earth crisis of 2010, after the New York Times broke a story that China had blocked exports of rare earth elements to Japan (Bradsher 2010).

Conspicuously absent from accounts of social and environmental impacts of rare earth mining in China was any sort of representation of the research and remediation efforts taking place in China before, during, and after this time. Although much of the research is published in Mandarin (Bai et al. 2001, 2004; Bai 2000), research on this topic has also been published in international Anglophone journals for decades (Bruce et al. 1963; Filipi et al. 2007; Hirano and Suzuki 1996). Since 2009, reportedly more than 3000 individual mining and processing enterprises in China had been closed in order to reduce severe pollution. This estimate includes all extractive and heavy industries, not just those within the rare earth supply chain. This was part of an initiative between the several government ministries and researchers at China Academy of Sciences, to assess the needs and formulate remediation plans for "Resource Exhausted Regions" (Guo et al. 2013; Wang 2011). This included a nation-wide effort to map, measure, and classify the different forms of soil and water pollution resulting from widespread resource extraction (Sodango et al. 2018), accompanied by closure and consolidation of heavily polluted industrial sites to reduce the number of mining and processing firms operating in the rare earth, cadmium, and coal sectors, to name a few (van der Kamp 2020). Implementation varied by region and included measures as diverse as training farmers to convert to growing crops that take up fewer contaminants from polluted soil to resettling entire cities. In the rare earth sector, these local transformations have been accompanied by a concerted effort to shift the country's status from a net rare earth exporter to a net importer as part of a broader risk mitigation effort. The campaigns have had results. In 2018, China became a net importer of rare earth ores for the first time since 1985.

Clean-up is expensive, but with the measurable immediate and projected long-term impacts of industrial pollution on everything from public health to agricultural productivity, it is the only option in a political context such as China where the abandonment of industrial sites and populations is not politically tenable (Baranovitch 2018; Klinger 2019; Yang and Ho 2018). The eventual need for an expensive clean-up effort was foreseen by planning and development officials, who, in author meetings and interviews undertaken 2009–2013, as well as in public pronouncements, repeated the axiom of: "grow first, clean up later" (L. Liu 2018) or "pollute first, control later" (Pang et al. 2019). This sensibility, long espoused by planners to the vexation of environmentalists and local communities, held that at some point China would be rich enough to focus on environmental conservation and well-being rather than economic development. This development doctrine has a long history, and was promoted in post-WWII debt-driven development doctrines enforced by Bretton Woods institutions and later by the Washington Consensus (e.g. Rich 1994). What is noteworthy is that, over the past decade, clean-up efforts have actually assumed increasing prominence in China's industrial areas. While this is good news for local residents and surrounding environs, it has entailed shifting primary production elsewhere, beyond China's borders.

The upshot is that for much of its history, the rare earth mining and processing industry has been able to evade sweeping innovations in social and environmental controls by instead utilizing the "spatial fix" (Harvey 1996) to relocate from areas of higher regulation and greater social and environmental protections to areas with fewer regulations and safeguards. The spatial fix has become standard practice with the elevation of cost reduction, deregulation, and profit maximization above all else under economic liberalism, a framework institutionalized globally by the General Agreement on Tariffs and Trade and later by the World Trade Organization. This is of course not unique to the rare earth sector. Similar historical trajectories through which industrial pollution is outsourced have been documented in other extractive, heavy, and manufacturing industries. This effectively buys more time for existing practices, which include (insufficient) pollution controls. Importantly, outsourcing pollution occurs not just across hemispheres (West to South) or countries (the US to China), but across sub-national regions as well (Eastern China to Western China; Eastern US to Southern/Western US).

This broader background is essential to understanding the lack of fundamental innovations in the manner in which we extract, produce, consume, and dispose of rare earth elements and other critical materials. Even with increasing measures and standards, waste-making and pollution prevails. Instead, the prevailing assumption is that we must pay for rare earth and other critical materials security by submitting to certain forms of social and environmental devastation—or more precisely, by assigning someone else's landscapes and lives to further social and environmental devastation, because the cost of mitigating social and environmental harms is presumed to be unthinkably high. Discourses of criticality, strategic importance, or supply chain instability preclude the possibility of taking adequate time to assess, plan, develop, and implement effective pollution controls. The question, then, is most often not about how to undertake necessary reforms to incentivize socially and environmentally sound forms of production, but instead, is about how to source rare earths as cheaply and efficiently as possible while minimizing political opposition in potential new sites of extraction and processing: a dynamic that has played out across the Americas, Central and South Eurasia, and with respect to the ocean floor and outer space. This is how the spatial fix operates: it is a decidedly antiquated, anti-innovative approach, one which is vested primarily in maintaining the late-twentieth century status quo characterizing global industrial geography, rather than finding substantive solutions to fundamental problems. Nevertheless, it is an approach to supply chain security that is fiercely defended, and it is fundamentally hostile to the kind of accounting that would generate the kind of quantitative data required to complete the analysis that initially motivated this chapter. The hostility is real: over the past decade, we have witnessed in the worldwide increase in assassinations of environmental defenders, the flagrant violations of limits on mining industries, and the rollback of individual rights and freedoms won in the post-WWII era in the interest of expedited extractivism (Scheidel et al. 2020).

Simultaneously, in the US context, the most common reasons given for the failure to substantially repatriate significant portions of the rare earth supply chain have basically come down to cost, which is often also an implicit criticism of "too much" regulation. Despite the geopolitically charged rhetoric about ending critical material dependence on China and regaining some measure of sovereignty over the supply chain via selective on-shoring, a refusal to deal squarely with the spatially and temporally expansive costs of mining and processing has hindered progress on this front. Without major policy interventions, it is impossible to beat the China price; it is too expensive for a firm to implement robust social and environmental safeguards; it is not cost-effective to source from non-Chinese sources; new firms cannot get up and running without being undercut from China; downstream firms may refuse to set up purchasing agreements with non-Chinese producers because it is not cost effective. But on the other hand, since 2011 the majority of legislative proposals to invest in rebuilding a modern and environmentally-sound rare earth supply chain in the US have largely failed to advance on the grounds that they constitute 'a government handout.' In this morass of truisms, two basic empirical questions are missing: how much does it cost to clean up? And how might this compare with the estimated costs of going about mining and processing the right way—which is to say clean, efficient, and socially and environmentally responsible—from the outset?

8.3 Scoping the Costs

While costs are frequently cited as an obstacle to doing things properly, it is important to note that the public always bears the cost of mining in some form: whether in any up-front price increases to offset the cost of investing in cleaner production practices, or in the tax-burdens imposed by clean-up and management, or in the multi-generational burden of lingering pollution. This fact should have the effect of shifting the conversation in a manner that makes the basic empirical research proposed herein possible: because it is the public that bears the tangible and intangible cost of more expensive or more scarce goods when supply chains are constricted, and it is the public that bears the cost of up-front investments in socially and environmentally sound production practices whether directly through increased costs passed on to the consumer or indirectly through publicly-funded research, basic information on these costs should be publicly available. But because such information is not (yet) available in a sufficient degree and reliability to inform rigorous quantitative analysis, all we can do here is offer an initial scope of the parameters that should guide such an analysis. If we set aside the concern with jurisdictional delimitation, an expansive scoping and preliminary assessment of the cost is possible. What emerges is a picture that takes the question of pollution control far beyond the mining site itself.

Whether preventing pollution beforehand or cleaning up afterward, interventions are required in many locations in order to mitigate the social and environmental harms of mining and processing. The sites of extraction, storage, transport, processing, and disposal each have surface, sub-surface, hydrological, and atmospheric dimensions, through which contaminants pass in the form of solids, liquids, and gases, and in large and small volumes. The manner in which contaminants interact with specific attributes of the physical environment (such as exposure to air or water, elevation profiles, wind and weather patterns, soil and groundwater properties, vegetation uptake or bioaccumulation among local fauna) must be considered in light of the formal and informal human interactions with specific features of the physical environment. Some populations are prone to prolonged exposure, such as front-line workers and communities located downstream and downwind. Their vulnerabilities differ from those of office workers, upstream and upwind communities (that may be more physically proximate), and occasional visitors. Individual vulnerabilities—and hence the bodily costs they might bear—vary by age, gender, and broader health profile. This raises the question of bench-marking. In short: what level of exposure to mining-generated pollutants counts as safe, and for whom? The precise thresholds of what constitutes acceptable exposure are often not disclosed to the public. The public instead must rely on official pronouncements of that existing levels of contaminants are either "safe" or "unsafe" without access to the criteria used to determine safety. However opaque, these thresholds are critical in determining the extent to which pollution must be controlled or cleaned up in the event of enforcement operations. Because these thresholds determine compliance and enforcement actions, they are often contested by industrial interests.

The designated (un)safe exposure thresholds are perhaps comparable between *ex ante* control and *ex post* clean-up, but the estimated costs of cleaning up after the fact are likely to be much higher because the spatial and temporal extent of the contamination is greater. Pollution, once discharged into the environment and taken up by soils, waters, plants, animals, and wind currents, is very difficult to contain (Bo et al. 2018). *Ex post* cleanup is much more complex because contaminants originating from the mining and processing operation might interact with pollutants from other sources to produce secondary pollutants (Bradley 1989). Furthermore, the effects of pollution exposure can occur far from the mining site, a fact which must be considered when delimiting the costs. The introduction of certain metals and materials into everyday industrial, logistical, and household applications constitutes a much more pervasive form of pollution occurring throughout the supply chain, which tends to be recognized only in isolation of the question of mining and processing. The effects of accumulated mining pollutants in the human body are not fully understood, neither in terms of their effects over the lifespan of the individual, nor their epigenetic effects over generations (Castillo et al. 2017; Li et al. 2019; Wu et al. 2007).

However, findings in the fields of major disease research, such as cancer, Alzheimer's and Parkinson's Disease, repeatedly affirm the importance of environmental contaminants (Chen and Ritz 2018; Yegambaram et al. 2015). Most existing research on the effects of accumulated mining pollutants in the human body focuses on lead, aluminum, and cadmium, with only limited research undertaken to understand the effects of exposure to contaminants specific to rare earth mines (Feng et al. 2007). Part of the limited scope of research is a matter of timing—the generations of children subject to the pervasive exposure to lead and aluminum via gasoline, housing materials, and cookware in the early and mid-twentieth century became, several decades later, senior citizens that disproportionately suffered from Parkinson's and Alzheimer's Disease. The individualized, household, and public burden of dealing with these diseases is, effectively, a massive *ex post* subsidy paid by individuals, households, and the public to a lead and aluminum industrial order of yesteryear. The emerging public health research question over the past decade with respect to rare earths is whether their increasing uses in electronics, glass and ceramics, and applications in fossil fuel refining and processing, and therefore their increasing incidence in emissions and atmospheric particulate matter may be transforming those of us who are here in the present into a mid- to late-twenty-first century variation of today's elder cohort of Parkinson's and Alzheimer's Disease patients.

Even if the future impacts manifest in different diseases, the example of Parkinson's Disease is instructive for the present as a rare earth-dependent industrial order has emerged, and with it, the increased intensity of exposure. In the US context, Kowal et al. (2013) estimated the 2010 cost of Parkinson's Disease to be several billion dollars: More than $14.4 billion allocated by the federal government, in addition to comparable medical expenses of $14 billion incurred by individuals with Parkinson's disease. Indirect costs were estimated at $6.3 billion, but are likely much greater because they primarily concerned the cost of reduced employment which came to $10,000 per person with Parkinson's disease and did not consider broader impacts on other members of the household, or of underpaid care workers. The upshot is that the cost of a single disease either resulting from or aggravated by exposure to environmental contaminants borne by individuals, families, and taxpayers is immense. In terms of dollars spent, it is orders of magnitude greater than building out an environmentally-responsible state-of-the-art industrial facility, or investing in materials science research to curtail the proliferation of these elements in our environments, the costs for which tend to be measured in millions, not billions, of dollars.

This example cites the costs for only 1 year of estimated expenditures in a single country and only for only one specific degenerative disease linked to lead and aluminum exposure. But the example is still instructive for defining and delimiting the costs of *ex post* vs. *ex ante* interventions. Of course, each contaminant behaves differently in human environments, leading to distinct effects and related costs. The profusion of lead, for example, into urban, residential, and industrial spaces in the early and mid-twentieth century generated a host of medical and political ills with immense costs attached. Considering these can help us think more clearly about the social and environmental costs associated with pollution from rare earth mining, processing, and proliferation. Lead has been associated with child development issues, increased aggression and risk-taking behavior, and therefore higher rates of violent crime, but it took decades to substantiate this link. These are human tolls borne by individuals, families, communities, as well as the criminal justice systems and education systems. In these cases, these tangible and intangible costs are extracted from the public well after the life of the mine, far from the mining sites at places of secondary manufacturing, application, and end-use, and long after potentially responsible parties might be identified to cover the monetary costs of damaging and degenerative diseases. In these cases, of course, it is not just the mining, processing, and manufacturing firms that are liable, but also the regulatory agencies that failed to uphold their charge to safeguard the public (Markowitz and Rosner 2013; Nevin 2000).

With respect to rare earth elements, preliminary research findings suggest that a comparably massive public subsidy, paid by the health of individuals, the care obligations of their families and communities, and public healthcare expenditures, is forthcoming to support the rare earth-dependent techno-industrial order of the present. These preliminary research findings largely concern the kinds of exposure that occur more generally, beyond the locations of mining sites. This is in part due to the fact that the studies are taking place in the Anglophone world, far from the largest or primary rare earth mining sites, which are located in China. Most studies published in English to date have been done on rats and selected human tissues; these seldom reference the epidemiological research carried out over the past three decades and published in Chinese scientific journals (Li et al. 2003; Hongzhong Liu 1996; Wang et al. 2002). More recent findings of exposure to rare earth elements include the cognitive impacts such as the impairment of spatial learning and memory (Huiying Liu et al. 2014), and reproductive impacts such as degraded sperm quality and accumulation in placental tissues (Kruger et al. 2010; Marzec-Wróblewska et al. 2015) in addition to long-standing findings of their neurotoxic, hepatoxic, and carcinogenic effects (Oliveira et al. 2014). Many of these effects result from the fact that the most commonly used rare earth elements are similar to calcium in terms of their size, bonding characteristics, coordination geometry, and donor atom preference, which means that they can displace calcium in some instances, and disrupt the activities of calcium and Ca-specific receptors and transmembrane channels (Kruger et al. 2010). This has widespread effects throughout the human body, and we do not yet understand how these effects may evolve over the human lifespan, much less what their broader sociological implications might be.

Therefore, these issues are larger than the question of cleaning up specific mining sites, but they are inseparable from the question of social and environmental impacts of rare earth mining and processing. These findings, and the subsequent questions they raise, highlight three main points. First, if effects are evident in human exposure that occurs far from mining sites, we can reasonably expect to see amplified effects for those who live and work in mining and processing areas. Second, the evidence of human exposure among those who never come within any reasonable proximity of mining and processing sites highlights the fact that social and environmental effects cannot be isolated to the site of production, processing, or disposal; they cannot be outsourced or sent a safe distance away if the elements mined are also made ubiquitous in our urban, industrial, and residential spaces in the form of finished goods or particulate. Third, because the social and environmental impacts generate long-term costs that are not geographically confined to mining sites, containment policies and jurisdictional approaches, which are, effectively, cost-containment approaches, they are not appropriate to meet the scope of the problem. Based on many forms of evidence marshalled from diverse approaches to studying the effects of exposure to pollution from mining, processing, and downstream activities, it is reasonable to conclude that an *ex post* approach produces these longer term social and environmental impacts, and therefore significantly increases the cost of clean-up and mitigation. But also, that an *ex ante* approach is unlikely to contain all pollutants, because the elements themselves can become pollutants at latter links in the supply chain. Nevertheless, a comprehensive quantitative comparison is needed. As the enduring bodily and environmental legacies of mining pollution show, there is effectively no way to fully remediate the health and environmental damages once they occur.

But that doesn't mean we should cease in all attempts to reduce harm. Even partial remediation and compensation is preferable to none among those whose landscapes and livelihoods have been directly damaged by mining and processing (e.g. Wang 2006; Zhang et al. 2001). Perhaps, then, the key to better research, policy, and practice is to stop pretending that any effort, no matter how extensive or expensive, can fully address remediation needs and instead focus on an expansive accounting of the costs. There is an uphill battle here, because the pressure on regulatory agencies to report effective enforcement activities, and the public-relations concerns of firms to minimize shortcomings in addressing environmental and health problems resulting from their activities, are counterposed to objective assessment and accounting.

When the sprawling and long-term costs generated by an *ex post* approach are compared to the estimated cost of *ex ante* controls, the choice seems clear. Similar to the *ex post* approach, there are many sites of intervention. But *ex ante* interventions at specific sites are more likely to limit the spatial and temporal extent of social and environmental harms generated by *ex post* approaches. While the up-front cost of *ex ante* controls is, quite logically, likely to be higher than the up-front cost of implementing fewer social and environmental safeguards, it likely is dwarfed by the sprawling direct and indirect, monetary and non-monetary costs generated by an *ex post* approach, the full effects of which are only beginning to be researched in the rare earth sector. In both *ex ante* and *ex post* approaches, it is important to note that exposure occurs not just at the sites of mining and production, but in urban, industrial, and logistical centers where particulates are dispersed via applications and end-uses of the element in question. Along these lines, greater production controls intended to limit the discharge of pollutants into environmental and biological systems might compel innovations in downstream applications, which could have the virtue of limiting the extent to which pollutants are released into our industrial, residential, and urban environments.

8.4 Analysis

The key cost differences are therefore two-fold: first, the costs of mitigating social and environmental impacts in an *ex ante* approach should, in theory, be more spatially and temporally contained to the sites of extraction and production, which suggests that the costs should therefore be more contained. But improvements also rely on investments among extensive networks of researchers, engineers, and industrial designers who must work in collaboration with similarly invested partners in government and industry to develop and implement meaningful *ex ante* controls and to improve them over time. These quite often federally-funded activities can involve significant cost outlays, but have a long record of generating net-positive returns (NRC 1995). Second, the costs of *ex ante* controls are more likely to be borne by a specific firm within a fixed timeframe, which may subsequently pass the cost on to consumers. This is in contrast to *ex post* approaches, which may impose the cost on the public indefinitely. While this is the optimal social and environmental outcome, it is not compatible with business competitiveness in our current political economic context. The additional up-front costs borne by a specific firm means that *ex ante* controls will likely increase the cost of production for the specific firm, which are either passed along to consumers or render the firm uncompetitive in a global free market context. The solution to this, then, is to provide funding or subsidies alongside strengthened mandates for social and environmental protections.

Whether as consumer, taxpayer, or pollution-affected individual, the public pays. The primary differences between the *ex ante* and *ex post* scenarios are the questions of: (1) who, specifically, has to pay; (2) what, precisely, the environmental impacts are, and; (3) how long the social and environmental impacts persist.

Ex post cleanup involves many more sites of intervention and a far greater spatial and temporal scope than *ex ante* controls, therefore it is likely that the *ex post* approach is not only much more expensive, but the foregoing analysis suggests that we are only partially aware of the full scope of social and environmental costs. This is important to incorporate into cost-benefit analyses, which are a long-standing mode of determining policy and investment decisions (Ackerman et al. 2005). The immediate implication is that the costs are most likely not properly accounted, which leads to stylized analyses that support an outcome that is preferred by a specific set of interests rather than one that is accurate to the situation. This question is important for whether and how rare earth supply chain concerns are resolved—locally and globally. Without a close look at the social and environmental costs across multiple spatial and temporal scopes, or a comparison between *ex ante* and *ex post* controls, it is all too easy to remain trapped in the status quo, punctuated by one-off policy responses. The current status quo is characterized by, on one hand, alarmism over dependence on China as well as the social and environmental devastation generated by the world's reliance on the country for the majority of rare earth processing and, on the other hand, whole or partial inaction to effectively address these challenges, due in significant part to an unexamined belief that the immense costs of proper social and environmental protection will hobble efforts to diversify rare earth supply chains from the outset. International scientific collaboration on pollution mitigation and clean-up in the rare earth sector, particularly with experienced researchers in China, would avoid duplication of efforts, enable a broader and more accurate accounting of the costs involved, and potentially shift the dynamic from one of bipolar antagonism to one of collaboration around shared concerns for cleaning up and securing rare earth supply chains in a manner that is sufficient to meet global demand.

8.5 Limitations and a Need for Greater Transparency

The preceding analysis is based on limited available data and extrapolates from a wide range of case studies. Costs associated with clean-up of specific mining operations—whether *ex ante* or *ex post*—are seldom disclosed or are only available in a manner that is highly fragmented and difficult to verify. The result is that rather than provide a straight-forward quantitative comparison between *ex ante* and *ex post* pollution controls, this chapter has instead scoped the basis for comparison and drawn on examples from other elements, the mining, processing, and proliferation of which are associated with specific diseases and social burdens. The paucity of available cost data suggests that cost-benefit analyses undertaken to guide policy and investment decisions are likely highly stylized, partial, and incomplete.

Further research efforts—and open publication of data and results—are needed in order to accurately assess the costs of rare earth mining and processing, to calculate the difference between *ex ante* and *ex post* pollution controls, and to guide effective policy formulation and implementation. Areas of critical research need include comprehensively identifying and characterizing the actual costs associated with cleaning up the rare earth supply chain, as well as investigations into the more spatially and temporally dispersed effects on the environment and public health that result from mining, processing, and the generalized introduction of rare earths in industrial, residential, urban, and logistical spaces.

References

Ackerman, F., L. Heinzerling, and R. Massey. 2005. Applying cost-benefit to past decisions: Was environmental protection ever a good idea? *Administrative Law Review* 57: 155–192.

Bai, L. 2000. Brief introduction to the existing environmental problems and comprehensive treatment of rare earth industry in Baotou [包头市稀土产业发展中的 环境保护问题及其治理整顿简报]. *Inner Mongolia Environmental Protection [内蒙古环境保护]* 12: 37–38.

Bai, L., Zhang, L., & Wang, L. 2001. *Radioactive environmental contamination and prevention measures in Baotou rare earth production*. Retrieved from Inner Mongolia Autonomous Region.

Bai, L., S. Wenli, and L. Zhong. 2004. The radiological impact of the Bayan obo rare earth and steel production on the surrounding environment. [白云鄂博矿在稀土和钢铁生产中放射性对周围环境的影响]. *Rare Earths* 25 (4): 75–77.

Baranovitch, N. 2018. The impact of environmental pollution on ethnic unrest in Xinjiang: A Uyghur perspective. *Modern China* 45 (5): 504–536.

Beckett, C., and A. Keeling. 2018. Rethinking remediation: Mine reclamation, environmental justice, and relations of care. *Local Environment: The International Journal of Justice and Sustainability* 24 (3): 216–230.

Bo, W., D. Xia, Y. Yu, H. Chen, and J. Jia. 2018. Source apportionment of soil contamination in Baotou City (North China) based on combined magnetic and geochemical approach. *Science of the Total Environment* 642: 95–104.

Bradley, S. 1989. Incorporation of metalliferous sediments from historic mining into river floodplains. *GeoJournal* 19 (1): 5–14.

Bradsher, K. 2010. Amid tension, China blocks crucial exports to Japan. *The New York Times*. Retrieved from http://www.nytimes.com/2010/09/24/business/global/24rare.html?pagewanted=all

Bruce, D.W., B.E. Heitbrink, and K.P. Dubois. 1963. The acute mammalian toxicity of rare earth nitrates and oxides. *Toxicology and Applied Pharmacology* 5 (6): 750–759.

Butler, K. 2012. Your Smartphone's dirty, radioactive secret. *Mother Jones(November/December),* n.p. Retrieved from http://www.motherjones.com/environment/2012/11/rare-earth-elements-iphone-malaysia

Castillo, S., P. Muñoz, M.I. Behrens, F. Diaz-Grez, and J. Segura-Aguilar. 2017. On the role of mining exposure in epigenetic effects in Parkinson's disease. *Neurotoxicity Research* 32: 172–174.

Chen, H., and B. Ritz. 2018. The search for environmental causes of Parkinson's disease: Moving forward. *Journal of Parkinson's Disease* 8 (1): S9–S17.

Cox, C., and J. Kynicky. 2018. The rapid evolution of speculative investment in the REE market before, during, and after the rare earth crisis of 2010–2012. *The Extractive Industries and Society* 5: 8–17.

EPA. 1997. *Cost of remediation at mine sites*. Retrieved from Washington DC:

Feng, L.X., X. He, H.Q. Xiao, Z.J. Li, F.L. Li, N.Q. Liu, et al. 2007. Ytterbium and trace element distribution in brain and organic tissues of offspring rats after prenatal and postnatal exposure to ytterbium. *Biological Trace Element Research* 117 (1–3): 89–104. Retrieved from http://download.springer.com/static/pdf/179/art%253A10.1007%252FBF02698086.pdf?auth66=1384038023_cfa65558bc7b4baa8576fc5cb90421c5&ext=.pdf.

Filipi, R., K. Nesmerak, M. Rucki, Z. Roth, I. Hanzlikova, and M. Tichy. 2007. Acute toxicity of rare earth elements and their compounds. *Chemicke Listy* 101 (10): 793–798.

Fisher, T., & Fitzsimmons, A. 2013. *Big wind's dirty little secret: Toxic lakes and radioactive waste*. Retrieved from http://instituteforenergyresearch.org/analysis/big-winds-dirty-little-secret-rare-earth-minerals/

Guo, W., R. Fu, R. Zhao, W. Zhao, J. Guo, and J. Zhang. 2013. Distribution characteristics and current situation of rare earth contamination of soils in the Bayan obo mining are and Baotou tailing reservoir in Inner Mongolia [内蒙古包头白云鄂博矿区及尾矿区周围土壤稀土污染现状和分布特征]. *Environmental Science [环境科学]* 35 (5): 1895–1900.

Harvey, D. 1996. *Justice, nature, and the geography of difference*. London: Blackwell.

Hirano, S., and K.T. Suzuki. 1996. Exposure, metabolism, and toxicity of rare earths and related compounds. *Environmental Health Perspectives* 104 (Suppl. 1): 85–95.

Jones, R. 2010. Inside China's secret toxic unobtainium mine. *Daily Mail*. Retrieved from http://www.dailymail.co.uk/news/article-1241872/EXCLUSIVE-Inside-Chinas-secret-toxic-unobtainium-mine.html

Klinger, J.M. 2017. *Rare earth frontiers: From terrestrial subsoils to lunar landscapes*. Ithaca, NY: Cornell University Press.

———. 2019. Environment, development, and security politics in the production of belt and road spaces. *Territory, Politics, Governance* 8: 657–675. https://doi.org/10.1080/21622671.2019.1582358.

Kowal, S.L., T.M. Dall, R. Chakrabarti, M.V. Storm, and A. Jain. 2013. The current and projected economic burden of Parkinson's disease in the United States. *Movement Disorders* 28 (3): 311–318.

Kruger, P.C., L.M. Schell, A.D. Stark, and P.H. Parsons. 2010. Lanthanide distribution in human placental tissue by membrane desolvation-ICP-MS. *Journal of Analytical Atomic Spectrometry* 25 (8): 1298–1307.

Lappi, P. 2018. Optimal clean-up of polluted sites. *Resource and Energy Economics* 54: 53–68.

Lawer, E.A., A.-C. Mupepele, and A.-M. Klein. 2019. Responses of small mammals to land restoration after mining. *Landscape Ecology* 34: 473–485.

Li, Y., X. Quan, D. Chen, L. Lin, and Z. Wang. 2003. The impact of endemic flourosis of children's intellectual development in Baotou. [包头市地方性氟中毒对病区儿童智力发育的影响]. *China Public Health Management* 19 (4): 337–338.

Li, S., M. Chen, Y. Li, and T.O. Tollefsbol. 2019. Prenatal epigentics diets play productive roles against environmental pollution. *Clinical Epigenetics 11* (82). https://doi.org/10.1186/s13148-019-0659-4.

Liebenberg, D., S. Claassens, and L. Van Rensburg. 2012. A multidisciplinary approach for the assessment of rehabilitation at asbestos mines in South Africa. *Environmental Earth Sciences* 67: 1237–1244.

Liu, H. 1996. A review of radioactivity in rare earth production and its effects on human health. [稀土生产的放射性及其对健康影响的概述]. *Rare Earths* 17 (4): 59–64.

Liu, L. 2018. A sustainability index with attention to environmental justice for eco-city classification and assessment. *Ecological Indicators* 85: 904–914.

Liu, H., J. Yang, Q. Liu, C. Jin, S. Wu, X. Lu, et al. 2014. Lanthanum chloride impairs spatial memory through ERK/MSK1 signaling pathway of hippocampus in rats. *Neurochemical Research* 39 (12): 2479–2491.

Margonelli, L. 2009. Clean energy's dirty little secret. *The Atlantic*. Retrieved from http://www.theatlantic.com/magazine/archive/2009/05/clean-energys-dirty-little-secret/307377/

Markowitz, G., and D. Rosner. 2013. *Deceit and denial: The deadly politics of industrial pollution*. Berkeley, CA: University of California Press.

Marzec-Wróblewska, U., P. Kamiński, P. Łakota, G. Ludwikowski, M. Szymański, K. Wasilow, et al. 2015. Determination of rare earth elements in human sperm and association with semen quality. *Archives of Environmental Contamination and Toxicology* 69 (2): 191–201.

Naidu, G., R. Seongchul, R. Thiruvenkatachari, Y. Choi, S. Jeong, and S. Vigneswaran. 2019. A critical review on remediation, reuse, and resource recovery from acid mine drainage. *Environmental Pollution* 247: 1110–1124.

Nevin, R. 2000. How lead exposure relates to temporal changes in IQ, violent crime, and unwed pregnancy. *Environmental Research* 38 (1): 1–22.

NRC. 1995. *Allocating Federal Funds for science and technology*. Washington, DC: National Academies Press.

Oliveira, M.S., I.M. Duarte, A.V. Paiva, S.N. Yunes, C.E. Almeida, R.C. Mattos, and P.N. Sarcinelli. 2014. The role of chemical interactions between thorium, cerium, and lanthanum in lymphocyte toxicity. *Archives of Environmental & Occupational Health* 69 (1): 40–45.

Owen, J., D. Kemp, É. Lèbre, K. Svobodova, and G. Murillo. 2020. Catastrophic tailings dam failures and disaster risk disclosure. *International Journal of Disaster Risk Reduction* 423: 1–10.

Pang, R., D. Zheng, M. Shi, and X. Zhang. 2019. Pollute first, control later? Exploring the economic threshold of effective environmental regulation in China's context. *Journal of Environmental Management* 248: 109275.

Pepper, M., M. Hughes, and Y. Haigh. 2021. Loophole or lifeline? The policy challenges of mines in care and maintenance. *The Extractive Industries and Society 8* (*3*). https://doi.org/10.1016/j.exis.2021.01.014.

Rich, B. 1994. *Mortgaging the earth: The World Bank, environmental impoverishment, and the crisis of development*. Boston: Beacon Press.
Scheidel, A., D. Del Bene, J. Liu, G. Navas, S. Mingorría, F. Demaria, et al. 2020. Environmental conflicts and defenders: A global overview. *Global Environmental Change* 63: 102104.
Shah, V., and A. Davery. 2020. Phytoremediation: A multidisciplinary approach to clean up heave metal contaminated soil. *Environmental Technology & Innovation* 18: 100774.
Sodango, T.H., X. Li, J. Sha, and Z. Bao. 2018. Review of spatial distribution, source, and extent of heavy metal pollution of soil in China: Impacts and mitigation approaches. *Journal of Health and Pollution* 8 (17): 53–70.
Sun, W., B. Ji, S.A. Khoso, H. Tang, R. Liu, L. Wang, and Y. Hu. 2018. An extensive review on restoration technologies for mining tailings. *Environmental Science and Pollution Research* 25: 33911–33925.
Unger, C., J.A. Everingham, and C.J. Bond. 2020. Transition or transformation: Shifting priorities and stakeholders in Australian mined land rehabilitation and closure. *Australasian Journal of Environmental Management 27* (1): 84–113. https://doi.org/10.1080/14486563.2020.1719440.
van der Kamp, D.S. 2020. Blunt force regulation and bureaucratic control: Understanding China's war on pollution. *Governance* 34: 191–209. https://doi.org/10.1111/gove.12485.
Wang, P. 2006. 'One in Seven,' Under the shadow of the tailings lake. *Eastern Broadcasting Network*. Retrieved from http://smgtv.eastday.com/eastday/node21/node148/node13584/userobject1ai161799.html
Wang, F. 2011. Analysis of common problems in resource-exhausted areas and the effect of public policy support [资源枯竭地区共性问题与公共政策扶持效应解析]. *Business and Economics* 2011 (17): 1–20.
Wang, L., W. Xing, and Y. Ji. 2002. A case-control study of lung cancer among workers in Baotou iron and steel company. *Journal of Baotou Medical College* 2002 (4): 10–14.
Wong, M., J. Wong, and A. Baker. 1999. *Remediation and management of degraded lands*. Routledge.
Wu, J., M.R. Basha, and N.H. Zawia. 2007. The environment, epigenetics, and amyloidogenesis. *Journal of Molecular Neuroscience* 34: 1–7.
Yang, X., and P. Ho. 2018. Conflict over mining in rural China: A comprehensive survey of intentions and strategies for environmental activism. *Sustainability* 10 (5): 1669.
Yegambaram, M., B. Manivannan, T.G. Beach, and R.U. Halden. 2015. Role of environmental contaminants in the etiology of Alzheimer's disease: A review. *Current Alzheimer Research* 12 (2): 116–146.
Zhang, L., W. Li, L. Yi, Z. Lihong, and Z. Yixiang. 2001. Inquest of residents in Baotou Bayan Obo Mining District 1995–1997. [1995~1997年包头市白云鄂博矿区居民死因调查]. *Journal of Baotou Medical College* 17 (1): 7–9.
Zhang, H., M. Zhang, Y. Wu, J. Tang, S. Cheng, Y. Wei, and Y. Liu. 2021. Risk sources quantitative appointment of ecological environment and human health in farmland soils: a case study on Jiuyuan District in China. *Environ Geochem Health* 43 (11): 4789–4803. https://doi.org/10.1007/s10653-021-00964-5.

Dr. Julie Michelle Klinger holds a PhD in Geography from the University of California, Berkeley. She is an Assistant Professor in the Department of Geography and Spatial Sciences at the University of Delaware; Associate Director of the *Minerals, Materials, and Society* program; principal investigator of National Science Foundation Award #2039857, *Characterizing the Global Illicit Trade in Energy-Critical Materials Using Machine Learning, Remote Sensing, and Multilingual Qualitative Research*; and a Co-PI on *Navigating the New Arctic: Sustainable Transitions Through Arctic Redevelopment.* She is a member of the International Standards Organization Technical Advisory Group 298: Rare Earth Supply Chain Transparency and Traceability. She has published numerous articles on rare earth elements, natural resource use, environmental politics, and outer space, including the award-winning book *Rare Earth Frontiers: From Terrestrial Subsoils to Lunar Landscapes.*

Chapter 9
The Social Conundrum of Eco-Centric Activism Against Oceanic Minerals

Saleem H. Ali

Abstract Extractive industries have caused immense social and environmental impact and yet remain the foundation of modern society. Minerals are the ultimate raw materials for human needs from fertilizers for food production to metals for infrastructure. Increasingly, minerals are also necessary for green technologies needed for climate change mitigation and adaptation. Yet, the opportunity to extract minerals on land is diminishing due to community opposition to new mines. Several countries such as El Salvador and Philippines have moratoria on new mining projects. Even during the presidency of Donald Trump, which has otherwise been receptive to extractive industries, a large metal mine in Alaska was rejected due to concerns about pollution impacts. In this context oceanic mineral resources are becoming an intriguing option since there are no direct social impacts in the deep sea. However, ecological impacts are largely unknown and remain a point of concern for conservationists. This chapter considers the ways in which international environmental governance of the extractive industries is developing to address such conflicts and how we can more effectively make decisions between social and environmental tradeoffs.

Keywords Deep sea mining · Industrial ecology · Polymetallic nodules · Life cycle analysis

9.1 Introduction: Divergence and Convergence of Social and Eco-Centric Activism

In March 2019, a working group of the International Commission on Stratigraphy which delineates such geological order for academia voted in favor of designating humanity's presence on the planet as an epoch. Twenty nine out of thirty four

S. H. Ali (✉)
Department of Geography & Spatial Sciences, University of Delaware, Newark, DE, USA
e-mail: saleem@udel.edu

S. Kalantzakos (ed.), *Critical Minerals, the Climate Crisis and the Tech Imperium*, Archimedes 65, https://doi.org/10.1007/978-3-031-25577-9_9

members of the "Anthropocene Working Group" supported the designation and voted in favor of starting the new epoch in the mid-twentieth century. They chose this date as the point when a rapidly rising human population accelerated the pace of industrial production, the use of agricultural chemicals, and other human activities. At the same time, the first atomic-bomb blasts embedded in sediments and glacial ice, thereby creating new minerals which would become part of the geologic record. A rather astounding set of analyses in this vein was conducted by researchers at the Weitzman Institute in Israel on the amount of physical matter on earth's surface highlights our impact. By their calculations in 2020, the total mass of all human-created material now exceeds all natural biomass (Bar-On et al. 2018). The same team also calculated that the 95% of the total biomass of mammals on earth was now either human or animals cultivated by humans for their consumption (Elhacham et al. 2020)! Human habitation has favored the ecological margin between land and the sea with approximately 40% of the current population of the planet estimated to live within 100 km of the oceans.

The oceans remain a vast but vulnerable frontier for scientific and industrial inquiry. Although often neglected as an immense sink for pollution and plunder, some of the earliest international environmental regulations also involved the oceans. The International Agreement for the Regulation of Whaling was signed in 1937 and is considered the first major multilateral environmental treaty (Susskind and Ali 2014). This agreement was further strengthened by the establishment of the International Whaling Commission and the formalization of the International Convention for the Regulation of Whaling in 1946. Despite the treaty having its ebbs and flows of national membership over the past several decades, the resilience and recovery of the world's largest mammals has been remarkable. A recent study of recovering whaling populations, published in the Proceedings of the Royal Society, showed that the humpback whale population in the southern oceans has gone from a low of 400 in the 1950s to over 25,000 in 2019.

Nevertheless, the Anthropocene continues to bring new challenges and opportunities for ocean conservation. An immediate threat which has garnered world attention with demonstrable impacts is the scourge of plastics pollution that poses to be a mounting menace for oceanic fauna. However, the environmental activist community is also stridently concerned about another potential longer-term threat – the advent of deep sea mining in coming years. A coalition of over eighty NGOs released a report in May 2020 on what was presented as a review of the scientific literature on the potential harm of seabed mining activity which is being considered under the United Nations Law of the Sea Convention (UNCLOS) (Oceans, Law of Sea). Earlier in the year, Fauna and Flora International published an assessment of their concerns to marine biodiversity from this activity. These reports are well-intentioned in their use of the precautionary principle and have received stellar endorsements from great doyens of oceanic conservation such as Sir David Attenborough and Sylvia Earle. However, precaution operates on a spectrum as with any human endeavor, and the principle cannot be used as an excuse for indefinite inertia in a world with competing challenges. Caution is in order, but indeterminate precaution is an untenable postulate that can lead to societal paralysis.

For functional purposes, a systems science approach is needed to consider the way forward as the deep sea mining enterprise has to try and optimize four overarching and potentially competing objectives:

(a) *Comparative environmental and social impact of mining on terrestrial versus oceanic ecosystems;*
(b) *Supply projections for key critical metals from terrestrial ore reserves (economically viable deposits) and resources (geologically available but currently not economically viable deposits); and from recycled sources and more resource efficient practices;*
(c) *Demand projections for critical metals needed for the transition to cleaner energy sources such as wind and solar power as well as for batteries for electric vehicles and smart-grids;*
(d) *Revenue generation potential for sponsoring Small Island Developing States (SIDS) which are partnering with industry for deposit exploration.*

The International Seabed Authority (ISA), which was formed over two decades ago as part of the United Nations Law of the Sea Convention, is planning to release a set of environmental standards and guidelines related to the implementation of its mining code in 2023. In this chapter, I provide a review of some of the key issues that can help to extricate the points of contention for more informed decision-making.

9.1.1 Comparative Ecological Impact Categories

There are a range of ecological impacts which are a point of conservation concern around deep sea mining and vary from the kind of location where mining is to occur. Mineral deposits in the deep seas can be found in (i) polymetallic nodules on the seabed; (ii) seamount metallic (ferromanganese) crusts; and (iii) sulfide deposits in extinct hydrothermal vents. Currently 18 exploration contracts have been granted for polymetallic nodules in the Clarion-Clipperton Fracture Zone (16), Central Indian Ocean Basin (1) and Western Pacific Ocean (1). There are seven contracts for exploration for polymetallic sulphides in the South West Indian Ridge, Central Indian Ridge and the Mid-Atlantic Ridge and five contracts for exploration for cobalt-rich crusts in the Western Pacific Ocean. Of these systems, the gravest concern of impact is near hydrothermal vent sulfide deposits. Notably the scaly-foot snail (*Chrysomallon squamiferum*), is found exclusively on three hydrothermal vent systems in the Indian Ocean at depths of over 2400 m and was placed on the International Union for Conservation of Nature (IUCN) Red List in June 2019 due to two of the three systems falling within the boundaries of exploratory mining licenses granted by the International Seabed Authority.

However, the most advanced mining exploration that has moved forward thus far and shown to be economically viable is largely focusing on polymetallic (Exploration Areas) nodules, which occur in arguably the least species diverse of the three ecosystems. No doubt, the ecology around these nodules is unique and all activities must proceed with great care regardless of absolute diversity estimates thus far. In particular, the microbes which are believed to have led to the accretion of the metallic deposits pose a fascinating area for research in terms of their pharmacological properties. To this end, the industry and the ISA have committed to only allowing for mining to occur if there are two "reference zones" which have been allocated for conservation purposes and for impact measurement purposes of at least equivalent size. Although some conservationists have questioned the efficacy of the concept of "no net loss", we undertake this approach constantly through protected areas management on land and in the oceans. A lot of the efficacy depends on the level of homogeneity of particular systems. Thus, coral reefs and active hydrothermal vents which are highly heterogeneous systems are not appropriate for such an approach, but the vast deep sea nodule plains may well be so. The characterization of these reference zones is an important protective feature that is now being studied in detail. Notably, an equivalent protective feature of this kind is not generally required in terrestrial mining sites that may occur in potentially even more biodiverse and vulnerable ecosystems such as rainforests.

Ultimately, the environmental impacts will need to be considered as a comparative life cycle assessment between terrestrial and oceanic minerals. Such a preliminary assessment has been undertaken by a team of industry-led scientists (including the co-founder of the Oceans Health Index, Dr. Steven Katona), and published in peer-reviewed journals with repute (Paulikas et al. 2020, 2022); lead author is multidisciplinary physicist and economist Daina Paulikas). Separate peer-reviewed studies have provided a framework for looking at the full impact considerations (Heinrich et al. 2020) which are likely to be less than terrestrial mining, largely owing to minimal processing energy requirements. Even more significantly, the waste generation and hazard of tailings dams (Global Tailings Review 2020) that come from terrestrial mines is mitigated in the case of oceanic minerals.

Assessments of the discharges from the shipping vessels that undertake initial processing of the ore residues, as well as the plumes of sediment generated along the water column as the minerals are retrieved to the surface are still to be fully evaluated. A team led by Thomas Peacock at MIT is undertaking some of this research for the International Seabed Authority. Studies from earlier deep sea disturbances that occurred in the 1970s and 80s during often clandestine exploration activities without any mitigation measures are not a suitable comparison (though can inform the standards development) with the current state of equipment and technology being tested. The same can be said of older techniques for terrestrial mining compared with modern practices.

Another area of concern that has been noted by some activists pertains to accidental methane releases from hydrate deposits in the deep sea which could contribute to carbon emissions. The temperature, pressure, and geochemistry of the region of the ocean where current mining is being considered so far does not appear to be a point of concern in this regard. Furthermore, methane extraction itself as a resource remains a point of consideration and if there was such a concern, it could potentially be harnessed as a resource, or for mitigation purposes.

The social impact comparisons of oceanic mining with terrestrial mining equivalents are far more clear in terms of evaluation. Most significantly, there are no major concerns of human rights abuses, child labor, human trafficking or property appropriations and resettlement that beset many terrestrial mining conflicts. Of course, the broader fishing and shipping industry has these social afflictions but the detailed oversight of all DSM vessels would prevent their prevalence in the sector. An area of potential social contention is the interaction between DSM activities and fisheries which could indirectly have social consequences. However, the single biggest threat to fisheries is commercial and often illegal fishing of vulnerable stocks. The impact of DSM on fisheries should certainly be monitored closely as the infrastructure is tested. There are also some interesting comparisons on fisheries in terms of noise and infrastructure development that can be compared with the oil and gas sector with which we have considerable long-term experience of measuring impacts. With oil and gas infrastructure there is far greater danger of catastrophic failure and large leaks or explosions from the deposit itself. However, the precautionary principle has been largely undermined in this context by major OECD countries for energy resource expediency.

Another comparison that is often raised with reference to social concerns about deep sea mining, is a comparison with the beleaguered Solwara 1 project by now defunct Nautilus Minerals. The demise of that project was due to a combination of investment flight and social concerns and opposition from coastal communities in Papua New Guinea, under whose maritime jurisdiction the project was being developed. None of those particular concerns would be applicable in remote parts of the ocean where deep sea mining is being developed. A plausible point raised by activists is that the environmental mitigation code being considered by the ISA a decade ago was to be modeled after the proposed measures from the International Marine Minerals Society (Code for Environmental Management). However, the current deliberations on the code are highly divergent and are more likely to consider best practices from the oil and gas sector as well as the maritime diamond mining sector operating off the coast of Namibia.

There is also a financial concern that is often noted about poor small island states investing in seabed mining and losing precious public funds in such ventures. For sure, caution is in order as there have been some heinous cases of poor extractive industry investments spurred on by misleading forecasts of fortune. However, there are numerous financial safeguards and wealth distribution mechanisms under the "common heritage of humankind" provisions of the Law of the Sea Convention in this regard. Furthermore, the minerals being extracted are the antithesis of fossil fuels in terms of their potential support for solving the climate crisis. Thus oceanic minerals have far greater chance for future market security than the conventional terrestrial stranded assets that are causing concern among development practitioners. Furthermore, the COVID crisis has led many small island states to consider ways of diversifying their economies (COVID-19), since reliance on tourism is so highly capricious in terms of extreme weather events or global emergencies such as the current pandemic. A country like Nauru, which has a tragic history of phosphate mining plunder on its territory, has invested in deep sea mining as a diversifying off shore investment, that would not have a direct ecological impact on its inhabitants. A detailed cost-benefit analysis of manganese nodules extraction for the Cook Islands conducted by the European Commission and the Pacific Community (SPC) suggested a net benefit of over $494 million over the life of a potentially feasible project.

9.1.2 *Mineral Demand and Supply Scenarios*

The potential for stranded investment concerns stems often from debates in the literature related to the future demand for specific oceanic metals such as cobalt, nickel and manganese. Commodity prices fluctuate wildly, and this can arguably have impact on the fortune of resource dependent economies, though strong institutions have shown to offset these effects. In any event, the latest demand scenarios indicate that substitution for the key metals that could be extracted from nodules is unlikely to occur because of fundamental electrochemical advantages of those elements and existing infrastructure investment in their manufacturing. Substitutes for copper have been predicted for decades and even after the advent of fiber optics cables that took a bite out of copper demand, the metal's need remains salient. Even if alternatives are found, the appliance or device intake for batteries is modular, and having a variety of products using different kinds of metal electrodes would be advantageous and lead to more resilience in the system.

Recycling of batteries and the potential for a circular economy are also important future supply sources for these metals. In China, "urban mining of metals" for recycling may well meet some aspects of demand. However, a range of scenarios studied by the World Bank in their most recent assessment indicate that recycling would not be able to meet demand needed for green technology uptake to reach the climate mitigation targets (Hund et al. 2020). Recycling also poses an intriguing sustainability challenge since in order to have stock for recycling it becomes necessary to make existing infrastructure more obsolescent. Overall, such an outcome may be worse in terms of sustainability metrics if there is too much focus on recycling primacy without a mix of supply sources. Furthermore, in terms of the development impact of the metals being harnessed by small island states, investors could even justify a premium price for consumers who want to support livelihoods in vulnerable states (similar to fair trade merchandise). Opportunities for metal stockpiling can also modulate price shocks and unlike agricultural products there is no "shelf life" challenge for long-term storage.

9.1.3 *Institutional Context and Errant Comparisons*

The prospect of a positive premium for deep sea mining metals seems very remote at present since it is instead being stigmatized by several civil society groups as a potential ecological threat. Unfortunately, the current polarization is a result of a lack of trust with institutions managing the process, particularly the International Seabed Authority. Criticism of the organization rests on its dual mandate to facilitate development while also regulating activities. There have been numerous proposals for improving the transparency of the organization's activities which deserve attention. In particular, the Legal and Technical Commission of the organization could be an area where further reform could help to build trust with environmental organizations and activists. Having greater independent governance and oversight of this committee could be considered within the framework of the UNCLOS mandate.

However, such reform, nor the need for further biological research, seem appropriate grounds to call for a moratorium on exploration licenses for deep sea mining. Comparisons with the moratorium under the Antarctic Treaty are not congruent since Antarctica has competing territorial claims and there is no corollary body of oversight and due diligence for mineral exploration similar to the ISA. Furthermore, there has in fact been a de facto moratorium on deep sea mining for the past 20 years even though technology had been developed for such activity going back at least 3 decades. The hiatus in development was largely owing to deliberative processes and research to address many of the concerns already raised and in the process of being addressed through the regulatory process. Any mining licenses, following viable reserves being found, would take several more years to develop in any case.

We are at a momentous time of the great energy transition alongside the need for diversification of oceanic economies of states. There are also international legal mechanisms for biodiversity protection which are emerging specifically for oceans, such as a new treaty for marine protection, alongside work by the High Level Panel on Oceans. Concerns over marine pollution from the surface activities of DSM would fall under the existing International Convention on the Prevention of Marine Pollution from Ships (MARPOL). There are broad conservation monitoring guidelines which have already been widely studied in recent literature, though the role which mineral exploration can play in resourcing such programs deserves greater attention.

9.1.4 The Way Forward for Sensible Conversations

Oceanic minerals development needs to be considered as part of a socio-ecological conversation about planetary sustainability at supply and demand nodes. Epithets such as "extractive capitalism" are also not appropriate for an activity which is in fact being undertaken through an unprecedented public-private partnership for global revenue sharing. Even though the activity will be far from any human habitation, there is still a presumption by some opponents of its impact being linked to social concerns on land. However, such discourse is holding the industry to a standard not even terrestrial mining is held to in terms of the locus of impact and legitimacy of community dissent. The level of global deliberations that are being carried out in this regard provide us with a rare opportunity to be *proactive* in environmental regulatory action rather than *reactive*.

The environmental movement should learn some of the lessons of its absolutist opposition to nuclear power which led to industrial research stagnation and a lost opportunity for climate mitigation because of extreme risk aversion. Resilience of natural maritime systems, even after nuclear exposure also needs to be considered. In the Bikini atoll where 23 nuclear tests were conducted (including the largest ever nuclear explosion), biodiversity studies conducted 50 years hence show remarkable recovery. Of course, the social consequences of this testing were immense but that is precisely why it is important to extricate social and ecological impacts in the high seas. Allowing for this frontier industry to cautiously proceed with careful monitoring and governance reform of the International Seabed Authority would be the sensible way forward for meeting the Sustainable Development Goal and targets related to the oceans (SDG 14, particularly Targets 14.7, A, B and C).

Civil society, industry and governments should also be willing to give each other some benefit of the doubt in this frontier area of human endeavor that by its very structure seeks such deliberative outcomes. Of course, there is pressure to develop prospects without undue delay due to investors having already put funds into particular ventures. Yet, such investment should be used as an opportunity to keep momentum in exploring the deep seas and harnessing resources where overall ecosystem integrity can be conserved. It is no wonder then that notable ocean conservationists, such as Greg Stone (former Chief Ocean Scientist for Conservation International), who have devoted their lives to the deep sea, are willing to consider this quest as a potential win-win opportunity that deserves a chance. Such professionals should not be blamed or shamed for engaging with industry or governments any more than environmentalists should be blamed for taking large donations from deep-pocketed conservation tycoons. Reasonable well-intentioned people can disagree about risk trade-offs and come to different conclusions and resource their endeavors accordingly. The Nobel Prize in Chemistry for 2019, for the invention of the lithium ion battery, shows us that good science can come from industry, government, and academia working together (the prize was shared by an industry scientist and two academics, one of whom had also worked for government). No one should make up their mind yet on the ultimate viability of this major planetary activity, but our debate should be based on science and due process of deliberative engagement with broader needs of society.

9.2 Applying Industrial Ecology as a Lens for Resolving the Conundrum

Applying industrial ecology methods to mining can be approached in two ways: (a) proximate resource usage in the production of the mine's output itself and (b) panoramic planetary resource implications of differentiated source streams for the mineral which requires an evaluation of the entire supply chain for the mineral. For the purposes of this paper, we will cover both facets of proximate and panoramic industrial ecology in the context of sourcing metals from the oceans. Indeed, the proximate impacts of the extraction process inevitably gets incorporated into systems-wide calculations. Four key ways in which we can apply industrial ecology approaches to the oceans are provided below:

(a) *Design for Environment:* As a starting point the design of industrial systems to consider ecological factors requires us to go back to key thermodynamic evaluation. The system should have fundamental attributes of mitigating energy and material usage in delivery; mitigate waste generation; and subsequently facilitate reuse and recycling of material inputs towards a circular economy. The extraction systems proposed for oceanic mineral deposits currently have the advantage of waste mitigation due to higher grade ores and the processing plants are designed to not require large tailings facilities that are needed for terrestrial metal ores.
(b) *Dematerialization:* Infrastructure needed to access mineral deposits in the case of oceanic deposits focuses on large seaborne vessels and deep underwater cables to transport the ore back to ships where there is initial processing of the ore, followed by transport to an onshore processing plant. This infrastructure needs to be compared in terms of material usage with the roads, railways and energy infrastructure needed when mines are developed on land. There is potential for dematerialization by moving extraction to the oceans.
(c) *Life Cycle Analysis (LCA):* A composite comparison between terrestrial and oceanic extractive industries from mines to markets can be most effectively carried out through an LCA. Much of industrial ecology discourse has developed around a range of LCA methodologies which range from calculating carbon footprints to social cohesion metrics. For oceanic minerals in the deep, the social LCA analysis comes out quite favorably overall since there are no communities to physically displace or property regimes to manage. For coastal oceanic mining the situation has more comparability to terrestrial mining.
(d) *Environmental impact analysis or assessment (EIA):* EIAs are similar to LCA in that they attempt to identify and quantify environmental impacts related to an activity. However, the scope and scale are different. Whereas life cycle analysis is concerned with "cradle-to-grave" or "cradle-to-cradle" energy and material flows within a product supply chain, environmental impact analyses are bound in time and space to a specific activity. In the context of mining, including deep sea mining, EIAs typically focus on the environmental impacts of a proposed project or series of projects within a defined geographic area. EIAs and LCAs can overlap, as in consideration of greenhouse gas emissions from mining operations but inform different actors and have different management implications.

Systems Synthesis: Much of the debate on oceanic minerals is also couched in terms of their use for green technologies such as wind and solar power infrastructure or electric car batteries. The imperative to build such infrastructure comes from concerns about human impacts on the planet's most critical life-sustaining indicators. The time sensitivity of the tipping points in this discourse on planetary boundaries makes the need for the technology development more urgent. Hence the option of waiting for alternative technology development and accumulation of recycled

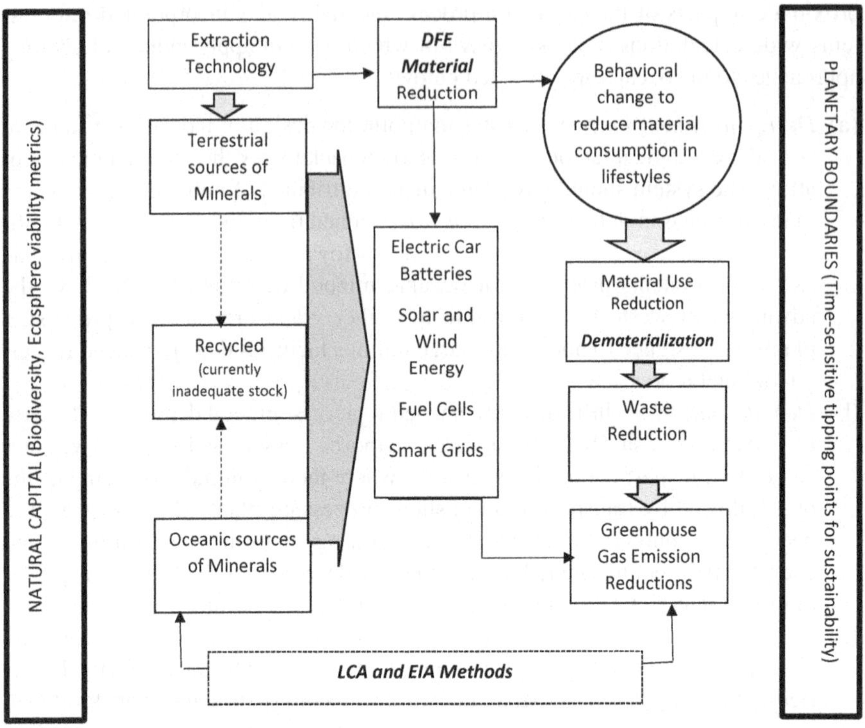

Fig. 9.1 Applying industrial ecology approaches to evaluating the viability and need of oceanic mineral extraction. The systems approach considers natural capital and key time-sensitive tipping points as the limiting "walled" parameters for decision-making. Industrial ecology methods are italicized

stocks of metals to be available for eventual circularity become less plausible. Recycled stocks accumulation is dependent on product durability whose benefits also need to be measured and compared with the need for more mined inputs through LCA. Climate change mitigation factors into such analyses as does oceanic ecology and thus the tradeoffs of sourcing metals to meet demand need to be considered. Within this context only enough metal so as to have a sustainable stock for recycling should be extracted, ultimately leading to a circular economy in the sector. The modularity of metal uses in electric car batteries makes such circularity plausible (Fig. 9.1).

9.3 Conclusion

The advent of the Anthropocene has posed the indelible impact of humanity on the planet but has also made the argument for conservation rather than preservation more contentious. There has been a history of concerns about environmentalists' view of preservation being at odds with the interests of indigenous human communities (Dowie 2011), as well as the dispossession of land by colonial settlers at the behest of conservationists (Spence 2000). Oceanic minerals extraction poses an opportunity for reduced direct social impact due to the absence of physical property rights, and the displacement of populations. Yet, activism against such mining is focused on uncertainty of impacts on biotic systems and their linkage to broader ecological sustainability of the oceans. Some activists have brought forth more intangible connections of ocean-faring communities to the deep sea as a social impact or the fisheries depletion prospect as a way to posit a social nexus. However, the proximity of impact differences in such cases is radically different from the kind of social impacts felt by communities in terms of child labor, direct pollution damage to property, and structural concerns of inequality and governance. A more useful way of social impact comparison would in fact be regarding the positive employment and state revenues for developing countries that terrestrial mining might provide in contrast with oceanic minerals.

The prospect of the use of these metals for green technologies has further complicated the decision-making process. Recycling, reduction in consumption, and organic alternatives are being presented as win-win outcomes. Yet there are thermodynamic tradeoffs between energy efficiency, safety, and cost which will make some metal demand from mined sources likely to continue. Recall, that despite numerous advances in technology, the lead-acid battery remains dominant in specific sectors. Similarly, cobalt, nickel, and manganese will remain significantly advantageous in key areas of mobile battery technologies even as alternatives such as iron phosphate are taken up by stationery battery sources. An industrial ecology approach provides us an opportunity to analyze and measure the social and ecological tradeoffs and provide a cogent way forward out of the conundrum of extractive consumption sources.

References

Bar-On, Y.M., R. Phillips, and R. Milo. 2018. The biomass distribution on earth. *Proceedings of the National Academy of Sciences* 115 (25): 6506–6511. https://doi.org/10.1073/pnas.1711842115.

Corson, Catherine, and Kenneth Iain MacDonald. 2012. Enclosing the global commons: The convention on biological diversity and green grabbing. *The Journal of Peasant Studies* 39 (2): 263–283. https://doi.org/10.1080/03066150.2012.664138.

Dowie, Mark. 2011. *Conservation refugees: The hundred-year conflict between global conservation and native peoples*. Cambridge, Mass: The MIT Press.

Elhacham, E., L. Ben-Uri, J. Grozovski, Y.M. Bar-On, and R. Milo. 2020. Global human-made mass exceeds all living biomass. *Nature* 588: 442–444. https://doi.org/10.1038/s41586-020-3010-5.

Global Tailings Review. 2020. Various publications online at https://globaltailingsreview.org/.

Heikkurinen, Pasi, Jenny Rinkinen, Timo Järvensivu, Kristoffer Wilén, and Toni Ruuska. 2016. Organising in the Anthropocene: An ontological outline for Ecocentric Theorising. *Journal of Cleaner Production* 113: 705–714. https://doi.org/10.1016/j.jclepro.2015.12.016.

Heinrich, Luise, Andrea Koschinsky, Till Markus, and Pradeep Singh. 2020. Quantifying the fuel consumption greenhouse gas emissions and air pollution of a potential commercial manganese nodule mining operation. *Marine Policy* 114 (April): 103678. https://doi.org/10.1016/j.marpol.2019.103678.

Hund, K., D. La Porta, T.P. Fabregas, T. Laing, and J. Drexhage. 2020. *Minerals for climate action*. Washington DC: World. Bank Publications Available at: https://pubdocs.worldbank.org/en/961711588875536384/Minerals-for-Climate-Action-The-Mineral-Intensity-of-the-Clean-Energy-Transition.pdf.

Paulikas, Daina, Katona Steven, Ilves Erika, and Saleem H. Ali. 2020. Life cycle climate change impacts of producing battery metals from land ores versus deep-sea polymetallic nodules. *Journal of Cleaner Production*. https://doi.org/10.1016/j.jclepro.2020.123822.

Paulikas, Daina, Katona Steven, Ilves Erika, and Saleem H. Ali. 2022. Deep-sea nodules versus land ores: A comparative systems analysis of mining and processing wastes for battery-metal supply chains. *Journal of Industrial Ecology* 26 (6): 2154–2177. https://doi.org/10.1111/jiec.13225.

Piccolo, John J., Haydn Washington, Helen Kopnina, and Bron Taylor. 2018. Why conservation scientists should re-embrace their Ecocentric roots. *Conservation Biology* 32 (4): 959–961. https://doi.org/10.1111/cobi.13067.

Rasche, Andreas, Georg Kell, and Ban Ki-moon. 2010. *The United Nations global compact: Achievements, Trends and Challenges*. Cambridge: Cambridge University Press. https://doi.org/10.1017/CBO9780511762642.

Spence, Mark David. 2000. *Dispossessing the wilderness: Indian removal and the making of the National Parks*. Revised ed. New York, NY: Oxford University Press.

Susskind, Lawrence, and Saleem H. Ali. 2014. *Environmental Diplomacy*. New York: Oxford University Press.

Saleem H. Ali holds the Blue and Gold Distinguished Professorship in Energy and the Environment at the University of Delaware and is also a Senior Fellow at Columbia University's Center on Sustainable Investment. Professor Ali has held the Chair in Sustainable Resources Development at the University of Queensland's Sustainable Minerals Institute in Brisbane, Australia (where he retains professorial affiliation). Previously, he was a Professor of Environmental Studies at the University of Vermont's Rubenstein School of Natural Resources where he was founding director of the Institute for Environmental Diplomacy and Security. His books include *Earthly Order: How Natural Laws Define Human Life* (Oxford University Press); *Treasures of the Earth: Need, Greed and a Sustainable Future* (Yale Univ. Press); *Environmental Diplomacy* (with Lawrence Susskind, Oxford University Press); *Mining, the Environment and Indigenous Development Conflicts* (University of Arizona Press); and *Islam and Education: Conflict and Conformity in Pakistan's Madrassas* (Oxford University Press). Corporate and government experience includes employment in General Electric's Technical Leadership Program, a Baker Foundation Fellowship at Harvard Business School, and a Research Internship at the UK House of Commons. He was chosen as a Young Global Leader by the World Economic Forum in 2011 and received an Emerging Explorer Award from the National Geographic Society in 2010. He is a member of the IUCN World Commission on Protected Areas and serves on the board of governors of Adventure Scientists and Mediators Beyond Borders. He is also a series co-editor for the University of Chicago Press on Environmental Science, Law and Policy. He received his doctorate in Environmental Planning from MIT, a Master's degree in Environmental Studies from Yale University, and Bachelor's degree in Chemistry from Tufts University (summa cum laude).

Chapter 10
The ESG Triangle: How Lithium Mining in Latin America Could Point the Way Toward Long-Term Environmental and Social Value Strategies

Owen Pell

Abstract Latin America's 'Lithium Triangle' accounts for over 60% of lithium reserves, creating commercial opportunities tied to lithium's role in electric vehicles and alternative energy storage, which the World Bank estimates may lead to an increase in lithium demand of 488% by 2050. Lithium extraction is concentrated in brine deposits and extracted in a water intensive process that affects both environmentally fragile salt flats and adjoining biodiverse wetlands—areas that also are linked, both economically and culturally, to Indigenous populations. Thus, economic benefits from mining are linked inextricably to long-term challenges of water depletion, usage and management, potential Indigenous population displacement, and reduced biodiversity. Managing these challenges now will be changing due to the recent Escazu Agreement (2018) and emerging ESG disclosure and diligence requirements, particularly from the European Union (EU), whose standards are the most developed. The Escazu Agreement guarantees rights to information, participation and access to justice in environmental matters, and mandates the involvement and protection of civil society organizations (CSOs) and personnel. Moreover, many companies in the lithium mining value chain or its financing will be subject to EU ESG standards requiring assessment and disclosure of how corporate activities affect, both positively *and* negatively, the environment and local communities as measured against EU ESG objectives. This chapter will show how the Escazu Agreement and EU ESG standards provide a case study for how companies can be encouraged to pay greater attention to long-term environmental and social issues, and the role that partnerships among industry, government focal points and CSOs could play in maximizing ESG value strategies.

O. Pell (✉)
The Auschwitz Institute for the Prevention of Genocide and Mass Atrocities and White & Case LLP, New York, NY, USA
e-mail: opell@whitecase.com

S. Kalantzakos (ed.), *Critical Minerals, the Climate Crisis and the Tech Imperium*, Archimedes 65, https://doi.org/10.1007/978-3-031-25577-9_10

Keywords Lithium triangle · ESG · Escazu agreement · European Union · Affected communities · FPIC · Water degradation

10.1 The Promise or Curse of a New El Dorado?

Lithium mining in Latin America has been called the "new quest for El Dorado" (Lopez-Calva 2022, 1). Lithium—sometimes called white gold—is at the center of a fast moving energy revolution that could dramatically reduce our dependence on fossil fuels, being used both to power electric devices and vehicles, and to store energy from renewable energy resources (Lopez-Calva 2022, 2–3). Accelerated efforts to decarbonize the economy because of the climate emergency have spotlighted lithium's importance as an input in the global economic transformation. It has, however, also given rise to scrutiny, not only over the way that lithium is being mined and the massive amounts of material that will be needed moving forward, but also across the entire supply chain from mine to market. Latin America's so-called Lithium Triangle, principally spread across deposits in Argentina, Bolivia, and Chile accounts for almost 60% of identified lithium reserves (Lopez-Calva 2022, 2–3). The speed of potential development and expansion could be significant, as global production has risen by over 20% since 2018 (USGS Mineral Commodity Summary 2022), and lithium production could by 2050 increase by up to 488% from 2018 levels (World Bank Group 2020, 73, 79, 103 (based on USGS data)). This scramble for lithium development, however, poses genuine environmental and social challenges.

Specifically, the drive toward rapid decarbonization that lithium can make possible also raises important questions relating to rights associated with the mineral itself, the water, land, and ecosystems where the mining occurs, and the land and cultural rights of the Indigenous populations that have lived in the mining areas for centuries. As noted by the World Bank, climate-smart mining will require mining companies and mining stakeholders to minimize the negative effects of mining so as to maximize the sustainability of mining activities (World Bank Group 2020, 17). Driving the long-term promise of the Lithium Triangle will be a different 'triangle' that will compel governments, companies, financial stakeholders, and civil society to understand and manage a complex and evolving collection of economic, environmental, and social and cultural opportunities and challenges. In the post-colonial era, stakeholders (and international conventions relating to environmental and social rights) demand that opportunities associated with lithium production, processing, and sales, which could power the green transition, also factor in the long-term environmental, social, and governance (ESG) costs of existing and expanding lithium mining. Defining this *ESG Triangle* are:

- The long-term environmental and social factors tied to lithium extraction within the Lithium Triangle. Lithium mining in Argentina, Bolivia and Chile raises specific environmental and social issues based on the mining method, the ecosystems in play, the lifespan of the mining process, and the history surrounding how governments and mining companies have engaged with local Indigenous populations who will experience the effects of the mining *and its aftermath*.

- The 2018 Regional Agreement on Access to Information, Public Participation and Justice in Environmental Matters in Latin America and the Caribbean (Escazu Agreement 2018). The Escazu Agreement guarantees the public rights to information, participation and access to justice as to environmental matters, and expressly guarantees both a role for *and the protection of* civil society organizations and personnel involved in environmental matters—especially with respect to vulnerable communities like the Indigenous peoples of the Lithium Triangle (Escazu Agreement 2018, Arts. 4–8, 2(a), 2(d) and 9).
- The disclosure, due diligence, and classification requirements tied to the EU ESG framework. The EU ESG standards go well beyond other regimes because they are mandatory, increasingly detailed, provide a formula for disclosure requiring companies to measure and balance claimed ESG benefits against actual or potential ESG detriments, and embrace the value of information in guiding investment and financial decisions. The EU ESG standards make no change to the duty of directors to maximize shareholder value, focusing instead on better defining how long-term value will be measured with respect to ESG objectives. With particular regard to lithium mining, the EU standards relating to assessing impacts on communities require companies to assess impacts on land rights, water availability and rights, Indigenous people's rights, and human rights defenders, and to ensure that local communities have free, prior, and informed consent in matters affecting them (EFRAG Draft ESRS S3—Affected Communities 2022e, 4–5, 7–8). The EU has stressed that investing in social objectives regarding impacted communities creates long-term value for companies and investors by reducing business risk and improving overall returns on investment (EU Platform on Sustainable Finance 2022, 22–25).

Although using long-term social and other non-financial data to assess business risks will be new to many companies implementing EU ESG standards, this has long been a part of World Bank analyses focusing on conflict risk and fragile states. For example, since 2009, in countries with World Bank development projects, almost 50% of conflict situations relate to natural resource degradation or depletion (World Bank Group 2021, viii, xii, 12–13). World Bank data also suggest that devolving resource rights to local communities can both reduce resource degradation and the risk of conflict (World Bank Group 2021, 12–13). Drawing on examples like this, and those involving peace building, conflict prevention and post-conflict situations, indicates that there are strategies for enhancing ESG value creation and reducing business risk with regard to lithium mining in Latin America. These strategies not only would address issues under the Escazu Agreement and EU ESG framework, but could also suggest best practices for addressing critical mineral mining in other locations.

10.2 The Environmental and Social Costs of White Gold

Lithium mining in the Lithium Triangle already involves significant costs, and further expansion to meet rapidly growing demand will involve further investments over the coming decades. Market participants will face a host of business risks that go beyond potential price volatility—volatility that may increase because over 60% of the value in the lithium value chain comes after the actual mining of the lithium and the recycling of lithium batteries may play an increasing role in the market. (Lopez-Calva 2022, 2; Purdy and Castillo 2022, 2–3, 22).

On the one hand, the Lithium Triangle contains almost 80% of the world's known lithium brine reserves (Marchegiani et al. 2019, 6). Accordingly, in a market of rapidly expanding demand, the Lithium Triangle is attracting outsized attention (World Bank 2021, 73, 79; Squire et al. 2020, 1; Marchegiani et al. 2019, 6–9). Lithium is extracted from three deposit types: brine, pegmatite (hard rock), and sedimentary deposits. Each process involves different extraction and processing methods. Brine and pegmatite currently provide most of the world's supply. Brine deposits currently are concentrated in Argentina, Bolivia, Chile, China and Tibet. Bolivian brine extraction comes with an additional cost challenge due to the lithium-manganese balance of the ore, which makes the refining process more costly (Marchegiani et al. 2019, 9; Kelly et al. 2021, 1–4).

On the other hand, the lithium in the Lithium Triangle is located in subterranean brine in desert or hyper-arid regions, where brine lies in watersheds under salt flats and associated desert wetlands, which are a home to many species of migratory birds and other flora and fauna and are populated by many Indigenous communities. Mining involves concentrating the brine by pumping large quantities of low concentration brine (which is 70% water) into large open-air evaporation ponds (Marchegiani et al. 2019, 9). The process is so water intensive that it is sometimes called "mining water" because of the enormous amounts pumped from the ground (Riofrancos 2022, 1). Data from Chile show that 500,000 gallons of water are required to yield a ton of lithium—raising the specter of a water sustainability challenge that may be endemic to this form of extraction (Berg 2021, 3–4; Molen 2022, 3). The brine, which is not usable for agricultural or domestic use, flows through a series of ponds over the course of a year or more, becoming steadily more concentrated—with some brine and other mineral salts leaching back into the watershed (Kelly et al. 2021, 3; Molen 2022, 1–3).

Environmental studies show that the aquifers found at the salt flats and wetlands in this region are fragile, especially if the aquifers do not connect to other aquifers, such that rainfall is the only means of recharge (Marchegiani et al. 2019, 35–36). Given that the Lithium Triangle is comprised of desert or hyper-arid areas with low rainfall rates, recharge cannot be assured. Brine extraction increases discharge from the aquifers and can affect freshwater levels in the watersheds in ways that rainfall may not replace. "These changes in water quantities could have profound effect on the salt flat ecosystem, affecting the flora and fauna" (Marchegiani et al. 2019, 36), which also would affect the agriculture and livestock activities of the local and

Indigenous populations. In addition, by disrupting the hydrological balance of the aquifers, brine extraction could cause brackish water to mix into the freshwater in the aquifers (Marchegiani et al. 2019, 35–37). Thus, the mining processes present an array of long-term issues that will change based on water usage, management, and remediation. These processes could adversely affect local Indigenous and other populations which rely on water for agricultural and domestic uses, including if climate change further reduces already low rainfall rates. As noted by the World Bank, 70% of the mining activities of the world's six largest mining companies are conducted in water-stressed areas, and "increasing extractive and processing activities *could* have serious environmental and social implications *if* these activities are not managed responsibly" (World Bank 2020, 31 (emphasis in original)).

Across Argentina, Bolivia, and Chile, reports are showing environmental effects tied to lithium mining. In Argentina and Chile, there are reports of aquifer depletion and residual water contamination, and in Argentina, Bolivia and Chile, fish and livestock damage has been linked to damage to local water supplies (Molen 2022, 2–4). In Chile, in response to depletion of water supplies used by local Indigenous populations, a regional water authority has declared that the local watershed is "exhausted" and is prohibiting new water rights from being allocated for lithium extraction from certain salt flats—a prohibition that does not apply to existing mining based on earlier water right grants (Riofrancos 2022, 2).

Beyond these reports, the greatest challenge relating to expanding mining in the Lithium Triangle appears to be the absence of detailed and specific environmental and hydrological studies of the various mining sites in this region. In particular, existing studies leave unclear what specific steps are being used to limit the impact of mining on affected communities in specific locations (Purdy and Castillo 2022, 22–23). There also has been very little research on the socio-economic impacts of lithium extraction at the local level of the affected communities (Agusdinata et al. 2018, 1; Marchegiani et al. 2019, 36–37). The paucity of locally-specific research raises additional questions. For example, are local Indigenous communities providing free, prior and informed consent (FPIC) to ongoing or expanding projects that will affect the environmental and human rights of those communities.

Argentina, Bolivia and Chile all have seen disputes and protests by Indigenous communities relating to development projects. These disputes relate to land and farming rights, access to watershed resources, the role of communities in the biodiversity and ecological balance of their regions, and access to the economic benefits of mining (Civicus 2020, 2022; Walter and Wagner 2021; Marchegiani et al. 2019, 26–27). In addition, Latin America has a long history of attacks on civil society defenders and local community representatives in disputes relating to mining and water issues, with nine Latin American states seeing attacks during 2021 (Global Witness 2021, 10–12). It should be noted, however, that beyond reports of threats and harassment, civil society defenders in the Lithium Triangle have not experienced the acute levels of violence seen in Brazil, Colombia, Mexico and Peru (see generally Global Witness Archive 2016–2020, showing that Argentina, Bolivia and Chile comprise less than 2% of the documented killings). Moreover, protests relating to potential mining operations and their possible effects on local communities in

Bolivia have stymied government efforts to advance several mining projects (Molen 2022; Squire et al. 2020).

In Argentina, there have been claims of barriers to participation, questions about how and where information is made available, and the limited time provided for communities to make FPIC decisions (Escosteguy et al. 2022, 1–3). In addition, there have been claims that mining companies have been slow to commit to publishing their plans for managing and remediating the long-term effects of their operations on water and environmental resources or on the communities those resources serve. Instead, companies appear to have focused on the short-term in making donations and investments in local communities. Although valuable, these short-term benefits do not address the long-term environmental, biodiversity, and social impacts of lithium extraction (Marchegiani et al. 2019, 26–28).

While this chapter does not argue for limiting or curtailing lithium mining, it highlights the recognized costs associated with lithium-related industrial development for which there must be a full, transparent, and ongoing accounting. As the World Bank acknowledges, addressing these costs is the key to maximizing the sustainability of mining (World Bank 2020, 17, 31)—something that is a key linchpin in the world's accelerated decarbonization transition in the face of climate change. As also recognized by the World Bank, the process by which local communities access, understand, and respond to information relating to the environmental and social costs of mining and other development projects is a central issue to sustainability. It is that process that is the focus of the Escazu Agreement, and why that agreement is important to lithium mining in Latin America.

10.3 Enabling Access and Building Civil Society Capacity: The Escazu Agreement

The 2018 Regional Agreement on Access to Information, Public Participation and Justice in Environmental Matters in Latin America and the Caribbean (known as the Escazu Agreement) applies to environmental matters (both public and private) in twenty-four states in Latin America and the Caribbean—including Argentina, Bolivia, and Chile. Its mandate is to make environmental matters more transparent, more open to effective public participation, more likely to support free, prior and informed consent (FPIC), and more likely to afford access to justice for affected local communities and with respect to environmental damage and remediation.

The impetus for the Escazu Agreement was the 1992 Rio Declaration. Specifically, Principle 10 of the Rio Declaration concerns public access to and participation in matters concerning the environment and communities, and Principle 22 recognizes that Indigenous people and their communities play a vital role in environmental management and development, and that states should enable those people and communities to participate in achieving sustainable development (Rio Declaration on Environment and Development 1992, 2–3, 4). Following the 2012 U.N. Conference on Sustainable Development (Rio + 20), ten Latin American and Caribbean states (including Chile) requested the UN Economic Commission for Latin America and the Caribbean (ECLAC) to assist in the drafting of an agreement based on the Rio Declaration Principles. Working with ECLAC, twenty-four nations drafted and negotiated the Escazu Agreement from 2012–2018 (ECLAC Escazu Agreement Implementation Guide 2022, 10–16).

The Escazu Agreement was considered so "ground-breaking" that the UN Secretary General hailed it as "a historic achievement" (Guterres 2021). Noting that this was the first environmental treaty for this region, the Secretary General highlighted the Escazu Agreement as "the first such treaty to include specific provisions for the protection and promotion of human rights defenders in environmental matters" (Guterres 2021). The Secretary General also emphasized that this agreement was integral to stepping up "efforts to curb the triple crisis of climate change, biodiversity collapse and pollution of the natural environment," and noted that the treaty was important as "a rights-based approach, fostering capacity-building and cooperation, and focusing on the most vulnerable" (Guterres 2021).

Although a milestone, it is important to understand that the Escazu Agreement did not emerge from a blank slate. Rather, the Escazu Agreement built on a strong foundation of international legal precedent relating to environmental rights and to the importance of civil society and civil society protection for the realization of human and environmental rights. These precedents are important because they show a broad consensus among governments—even governments that stand to gain a great deal from business activities like mineral extraction—that sustainable development is more valuable than business activities that harm the environment and communities, and that reducing risks associated with non-sustainable development will yield better value and results in the long-term. These themes are central to the Escazu Agreement, and are at the heart of the EU ESG framework discussed below.

The first relevant precedent for the Escazu Agreement was the 1988 Additional Protocol to the American Convention on Human Rights which expressly provides that all states "shall promote the protection, preservation, and improvement of the environment." In addition, the Protocol provides that "[e]veryone shall have the right to live in a healthy environment" (Additional Protocol 1988, Art. 11). More specifically, however, the Escazu Agreement then built on two other documents addressing how rights like environmental rights are best realized.

First, the 1998 Convention on Access to Information, Public Participation in Decision-Making and Access to Justice in Environmental Matters began developing what participation, information access and access to justice meant in environmental matters (Aarhus Convention 1998). These provisions also recognized the importance of civil society organizations (CSOs), and that those exercising their right to participate "shall not be penalized, persecuted or harassed in any way for their involvement" (Aarhus Convention 1998, Art. 3.8). Second, the 1999 U.N. General Assembly Resolution on Human Rights Defenders, expressly recognized the importance of CSOs and CSO enabling environments (UN Resolution on Human Rights Defenders 1999, Arts. 13, 16, 18). That resolution also provides that states "shall take all necessary measures" to protect anyone, including CSOs and their personnel from "any violence, threats, retaliation, *de facto* or *de jure* adverse discrimination, pressure or any other arbitrary action as a consequence" of their exercising their rights of participation (UN Resolution on Human Rights Defenders 1999, Arts. 12). The Aarhus Convention and UN Resolution on Human Rights Defenders set the stage for the Escazu Agreement to further delineate the role to be played by CSOs in the process of realizing environmental rights to information, participation and access to justice.

Another important precedent for the Escazu Agreement is the 2017 Advisory Opinion of the Inter-American Court of Human Rights which comprehensively addresses the relationship of environmental and human rights. This ruling was handed down while the Escazu Agreement was being negotiated, but provided extensive legal support for the duties ultimately delineated in the treaty.

The Court stated that, based on a broad collection of international and regional treaties, there is a "close connection between environmental protection, sustainable development and human rights" (Advisory Op. 2017, 24). The Court also recognized that the relationship between a healthy environment and the protection of human rights also relates to the territorial rights of Indigenous peoples (Advisory Op. 2017, 21, 28). The Court made clear that states have an "obligation to prevent" environmental harm and degradation (with the environment defined broadly to include land, water, atmosphere, and flora and fauna), and an obligation giving rise to duties to regulate and monitor the environment, to require environmental impact assessments, to prepare environmental contingency plans, and to mitigate environmental damage (Advisory Op. 2017, 51–68).

Finally, based on express provisions in the American Convention on Human Rights and other international agreements, the Court stated that the realization of environmental rights *requires* states to provide access to environmental information, opportunities for public participation in environmental decision-making, and access to justice in support of state duties relating to environmental rights (Advisory Op. 2017, 81–89). Regarding Indigenous peoples, the Court went further, saying that states "must ensure the rights to consultation and to participation at all stages of the planning and implementation of a project or measure that could have an impact on the territory of an indigenous or tribal community" (Advisory Op. 2017, 88) The Court also recognized FPIC principles, stating that

> in addition to receiving and providing information, the State must make sure that members of the community are aware of the possible risks [of a project or measure], including health and environmental risks, so that they can provide a voluntary and informed opinion about any project that could have an impact on their territory within the consultation process. (Advisory Op. 2017, 88; principles recognized in Article 19 of the Universal Declaration on the Rights of Indigenous Peoples 2007).

The Advisory Opinion is important because it collects relevant international legal precedents and places them into the context of environmental and human rights in Latin America and the Caribbean. This set the Escazu Agreement on even firmer legal footing, even though it was the first treaty of its kind in this region. The Escazu Agreement closely tracks the duties delineated in the 2017 Advisory Opinion. The treaty, however, is not self-executing, requiring states to enact legislation to ensure compliance. Nonetheless, the treaty creates express duties that should be enforceable under the American Convention on Human Rights, and potentially under the laws of other nations that allow for claims arising under international law.

The treaty requires states to "guarantee the right of every person to live in a healthy environment and any other universally-recognized human rights related to the" treaty {Escazu Agreement 2018, Art. 4(1)}, which thereby reaches social and human rights issues arising in environmental matters. The treaty creates broad obligations for states to develop and provide access to information relating to environmental matters, and mandates that states ensure access to justice in reviewing environmental decisions and in providing remedies for environmental harms and damages (Escazu Agreement 2018, Arts. 5.3, 7.2–7.8, 8.2–8.6). Significantly, states also must designate "impartial entities or institutions" with "autonomy and independence to promote transparency in access to environmental information" (Escazu Agreement 2018, Art. 5.18). In terms of information dissemination, states must develop information systems to track pollution and waste release relating to air, water, soil and subsoil, and must disseminate information relating to imminent environmental harms (Escazu Agreement 2018, Arts. 6.3–6.5).

The Escazu Agreement then concentrates on making access to information and justice more meaningful by focusing on public participation in environmental matters. The treaty defines "access rights" to include "the right of public participation in the environmental decision-making process and the right to access to justice in environmental matters" (Escazu Agreement 2018, Art. 2(a)). The "public" *expressly includes* civil society organizations (CSOs)—i.e., "one or more natural or legal persons and the associations, organizations or groups established by those persons, that are nationals or that are subject to the national jurisdiction of the State Party" {Escazu Agreement 2018, Art. 2(d)}. The treaty then focuses on civil society organizations (CSOs) in seeking to build capacity within society to make access rights more likely to succeed.

The treaty expressly provides that each state party "shall guarantee *an enabling environment* for the work of persons, associations, organizations and groups that promote environmental protection, by recognizing and protecting them" {Escazu Agreement 2018, Art. 4.6 (emphasis added)}. Underscoring the importance of CSOs, the agreement also recognizes that some communities will face challenges in benefiting from the treaty. "Vulnerable" persons are "those persons or groups that face particular difficulties in fully exercising the access rights recognized in the [treaty], because of circumstances or conditions identified within each [State] Party's national context" (Escazu Agreement 2018, Art. 2(e)). States are required to "facilitate access by persons or groups in vulnerable situations" (Escazu Agreement 2018, Arts. 5.3–5.4, 6.6). That access must include rights to challenge and appeal decisions relating to environmental matters (Escazu Agreement 2018, Art. 5.2(c)).

To ensure greater transparency, public participation, and access to justice, including for vulnerable communities, the Escazu Agreement then takes the unprecedented step of expressly acknowledging the importance of active CSO engagement by mandating that states *guarantee* that CSOs can exist while also guaranteeing the safety of CSOs and their personnel. Thus, states "shall guarantee a safe and enabling environment for persons, groups and organizations that promote and defend human rights in environmental matters, so that they are able to act free from threat, restriction and insecurity" (Escazu Agreement 2018, Art. 9.1). With respect to attacks on CSOs, the treaty mandates that states

> shall take adequate and effective measures to recognize, protect and promote all the rights of human rights defenders in environmental matters, including their right to life, personal integrity, freedom of opinion and expression, peaceful assembly and association, free movement, as well as their ability to exercise their access rights … (Escazu Agreement 2018, Art. 9.2).

Moreover, each state "shall also take appropriate, effective and timely measures to prevent, investigate and punish attacks, threats or intimidation that human rights defenders in environmental matters may suffer while exercising the rights set out in the [treaty]" (Escazu Agreement 2018, Art. 9.3). These CSO-related provisions of the Escazu Agreement were part of the preliminary negotiating draft prepared by ECLAC, and were specifically based on the Aarhus Convention, the UN Resolution on Human Rights Defenders, and examples of national laws already in place in Latin America and the Caribbean (ECLAC Preliminary Document 2015, Arts. 5 & n. 49, 9 & n. 149).

The Escazu Agreement's focus on CSOs follows from the Aarhus Convention, which recognized that CSOs "have an important role to play in contributing to making the public more aware of questions relating to all human rights and fundamental freedoms" (Aarhus Convention 1998, Art. 16). CSOs also play a role in "safeguarding democracy, promoting human rights and fundamental freedoms and contributing to the promotion and advancement of democratic societies, institutions and processes (Aarhus Convention 1998, Art. 18). This is consistent with evidence of CSOs helping to achieve better human rights and access to justice outcomes, including their ability to assist governments and industry in building transparency, gathering data on the needs of vulnerable communities, monitoring impacts of government and business activity, and devising solutions to sustainable development challenges (Task Force on Justice 2019, 22–24, 30–42, 44–58, 60–90).

As noted above, given the importance of the FPIC process generally, and the presence of vulnerable and Indigenous populations in the Lithium Triangle, the Escazu Agreement creates an important guidepost for companies engaged in ongoing or expanding lithium extraction. It places a premium on information access, informed public participation, and access to justice and remedies regarding environmental issues. The Escazu Agreement also encourages companies to focus on local communities and on CSOs working with them, something the World Bank has observed mining companies should embrace in making mining more sustainable (World Bank Group 2020, 31). These engagement processes will take on added importance in light of the emerging EU ESG framework.

10.4 The Emerging EU ESG Framework and Long-Term Value Creation

The role of the corporation in society has evolved over the last 75 years—from the inception of corporate philanthropy, to the concept of corporate social responsibility, and finally to prescribed rules for ESG investing today. Over time, there has also been a trend from voluntary reporting to mandatory disclosure and due diligence requirements, and the pace of change has been accelerating. In 2011, the UN Guiding Principles on Business and Human Rights were endorsed, paving the way for a global standard for preventing and addressing the risk of adverse human rights impacts linked to business operations (UN OHCHR 2011). Leading international financial institutions like the IFC and OECD then incorporated these principles into their lending principles and guidance for responsible business conduct (IFC 2012; OECD 2011). It became increasingly clear, however, that voluntary standards would not address key issues relevant to the challenges of Climate Change and sustainable development. The EU recognized this over 20 years ago, beginning as early as 2001 to implement binding disclosure, accounting, and data collection standards relating to Climate Change and other environmental issues (see generally Eur. Parliament Proposed Directive 2022, 4–8 nn. 14–32 (reviewing EU directives and regulations dating back to 2001)). The EU, moreover, also was the driving force behind the Aarhus Convention, the first international convention focusing on the role and protection of civil society in furthering environmental rights, and access to information, participation, and justice in environmental matters. The EU and over twenty-five EU members are signatories of the Convention (Aarhus Convention 1998).

Governments, most notably in the EU, now recognize that they cannot mobilize the capital required, nor are they equipped to make or manage the myriad economic and other decisions required to solve climate change challenges. In response, the EU is focusing on making responsible business conduct an integral part of the value maximization duty that directors and executives already owe to shareholders. Through ESGs, the EU is relying on markets—and disclosure-based market and pricing decisions—to address the remarkable complexities the world faces. In taking this course, the EU has established itself as the global leader in ESG business practices relating to the fight against climate change and for sustainable development.

As explained by the European Commission, the EU ESG approach goes to directors' "duty of care to promote the interests of the company and pursue its objectives."

> The initiative aims to ensure that sustainability is further embedded into the corporate governance framework with a view to align better the long-term interests of management, shareholders, stakeholders and society. It aims at improving the framework to incentivise [sic] corporate boards to integrate properly stakeholder interests, sustainability risks, dependencies, opportunities and adverse impacts into strategies, decisions and oversight. It would serve the following specific objectives: help companies' directors to establish longer-term time horizons in corporate decision-making and withstand short-term pressures, strengthen the resilience and long-term performance of companies through sustainable business models and help reducing adverse impacts. It would create legal certainty and level

> playing field as to the necessary measures to be taken by companies to identify, assess and mitigate adverse impacts in the value chain. (Eur. Comm'n 2020, 3).

At present, the EU ESG framework includes:

- Non-Financial Reporting Directive (NFRD) covers public companies with over 500 employees, financial institutions, and insurance companies. Largely comprehensively replaced by the Corporate Sustainability Reporting Directive—defined below—requiring companies to use non-financial data in how they report both on how sustainability issues affect their performance, position and development, and on their impact on people and the environment (Eur. Parliament Directive 2014).
- Sustainable Finance Disclosure Regulation (SFDR) imposes mandatory obligations for asset managers and other financial market participants to provide standardized disclosures for integrating ESG factors at both an entity and product level. The SFDR contains important definitions relating to sustainability, and disclosures require due diligence and measurement of good governance practices (Eur. Parliament Regulation 2019). As used throughout the EU ESG legislation, "due diligence" in this context is not limited to what a company may do before it enters into a transaction with another company; rather, what is meant here is the inward focusing and ongoing process by which a company analyzes itself and its value chain as to ESG issues (EFRAG Draft ESRS 1—General Requirements 2022b, 13–15).
- Corporate Sustainability Reporting Directive (CSRD) (taking effect 1 January 2024 for reporting fiscal year 2023), would replace NFRD and apply to any companies (listed or not) with €20 million in assets, €40 million in revenue and/or 250 employees in the EU. The CSRD revises and strengthens existing rules, and extends reporting and disclosure requirements. It also imposes ESG due diligence mechanisms, and provides for detailed reporting standards to be developed by the European Reporting Lab at the European Financial Reporting Advisory Group (EFRAG) (Eur. Parliament Proposed Directive 2021). The European Parliament adopted a near-final version of the CSRD as of November 2022 (Eur. Parliament Press Release November 2022).
- EU Taxonomy Regulation creates a common classification system and language that investors can use when investing in projects and economic activities that claim to have a substantial positive impact on the environment—a requirement that the CSRD extended to apply to social and governance objectives (Eur. Parliament Regulation 2020).
- EU Commission Proposal for a Directive on Corporate Sustainability Due Diligence (published February 2022; intended entry into force 2024) would require, as part of due diligence, a duty to identify, monitor, and take steps to remedy actual, and prevent or mitigate potential, adverse impacts on human rights in a company's operations, and those of subsidiaries and the value chain (Eur. Parliament Proposed Directive 2022). This proposal coincides with France, Germany, and Norway passing national laws on required due diligence, and

pending proposals or initiatives in Belgium, the Netherlands, and Austria, Denmark, Luxembourg, Spain, Switzerland, and Sweden, respectively.

Although none of the largest lithium mining companies is EU-headquartered, it is highly likely that the EU ESG framework nonetheless will affect mining in the Lithium Triangle because, as noted above, the CSRD lowers the EU thresholds to cover companies (listed or not) with €20 million in assets, €40 million in revenue and/or 250 employees in the EU. In addition to mining companies (which may have substantial EU processing, distribution, or trading operations) and companies using lithium batteries (e.g., tech and automobile companies with EU operations), the EU ESG framework also will reach major banks, investment banks, insurance companies, pension funds, and fund managers located in or with substantial EU operations.

10.4.1 Assessing and Disclosing ESG Risks and Rewards

Given the importance of lithium production to the green transition, companies and investors perhaps take for granted that investing in lithium extraction must be an ESG-positive activity. Importantly, under the EU ESG framework, this is not necessarily true, and that will create important disclosure and business strategy issues for companies and investors.

Under the SFDR, "sustainability" includes environmental, social and governance objectives (as developed by EFRAG) which are used to assess corporate activities, factors or events (Eur. Parliament Regulation 2019, Art. 2(17), 2(22), 2(24)). Under the Taxonomy, a sustainable investment must demonstrably contribute to achieving an ESG objective (Eur. Parliament Regulation 2020, Art. 9). Under the Taxonomy, "sustainable land use and management" includes "adequate protection of soil biodiversity, land degradation neutrality and the remediation of contaminated sites." (Id., Art. 15) Also, environmental impact assessments must be made throughout the full life cycle of a product or service—through product/service end of life. (Id., Art. 17(2))

There are five environmental objectives: (i) climate change (which includes mitigation and adaptation); (ii) transition to the circular economy; (iii) pollution prevention and control; (iv) sustainable use and protection of water and marine resources; and (v) protection and restoration of biodiversity and ecosystems (Eur. Parliament Regulation 2020, Art. 9). The CSRD then expands this to include Social and Governance Objectives (Eur. Parliament Proposed Directive 2021, 79–82). Although awaiting final approval, draft disclosure and due diligence standards for environmental, social and governance objectives were submitted by EFRAG to the European Commission in November 2022 as part of the Platform on Sustainable Finance established under the Taxonomy Regulation (EFRAG Draft ESRS Cover Letter 2022a; Eur. Parliament Regulation 2020, Art. 20; EU Platform on Sustainable Finance 2022, 10).

Social objectives require respect for recognized human rights by requiring that companies assess impacts on workers, consumers, and communities affected by a company's activities (including a company's value chain). Governance objectives include, among other things, assessments and disclosures on a company's (i) corporate culture and business conduct policies; (ii) political influence and lobbying activities; and (iii) anti-corruption/bribery systems. (EFRAG Draft ESRS G1—Business Conduct 2022d; see Eur. Parliament Proposed Directive 2021, 79–82). Finally, companies must assess how their activities "promote adequate living conditions for all," including (i) by fulfilling basic human needs; (ii) providing access to essential economic services and infrastructure; (iii) by creating decent employment; and (iv) by promoting peaceful and inclusive societies (EU Platform on Sustainable Finance 2022, 32–38). The drafters of the EFRAG disclosure rules stressed that investing in Social and Governance Objectives will create opportunities for investors, including by reducing various forms of business risk. This includes data showing that returns on investment are higher in societies with more rights and stable, transparent institutions (EU Platform on Sustainable Finance 2022, 23–25).

With particular importance to the Lithium Triangle, the EFRAG disclosure rules on affected communities require companies to assess and disclose how they will identify and manage the impact of their operations. Impacts must be assessed on, among other things, (i) land and water rights; (ii) the rights of Indigenous peoples; (iii) human-rights defenders; and (iv) the accessibility and availability of basic economic infrastructure and services (EFRAG Draft ESRS S3—Affected Communities 2022e, 4–6). In addition, as to Indigenous peoples free, prior and informed consent (FPIC) rights are highlighted, and generally, the process by which companies engage with community stakeholders also must be disclosed (EFRAG Draft ESRS S3—Affected Communities 2022e, 7–9; EU Platform on Sustainable Finance 2022, 37). The required disclosures call for real detail, including how companies engaged with communities, with whom and how often they engaged, and which senior executives have "operational responsibility" for the engagement. As to Indigenous peoples, companies "shall disclose whether and how indigenous peoples have been consulted on the modes and parameters" of the engagement itself (EFRAG Draft ESRS S3—Affected Communities 2022e, 7–8). Thus, affected communities "are a key group of affected stakeholders" under the disclosure rules (EFRAG Draft ESRS S3—Affected Communities 2022e, 5).

The Taxonomy then sets forth a formula, of sorts, for understanding how these Objectives factor into corporate disclosure, due diligence, and strategic planning. This formula will be important to any company attempting to assert that a particular activity is ESG-positive. A company can assert that an activity is ESG-positive, i.e., a "sustainable economic activity" *only* if that activity (i) contributes to at least one stated objective, *and* (ii) will do no significant harm (the "DNSH" principle) to any other objective (Eur. Parliament Regulation 2020, Arts. 9, 17). As to social and governance objectives, a sustainable economic activity *also* must meet "minimum safeguards" as expressly delineated in the OECD Guidelines for Multinational Enterprises, and UN Guiding Principles on Business and Human Rights (cf. Eur. Parliament Regulation 2020, Art. 18; see also EFRAG Draft ESRS Explanatory

Note 2022f, 11 discussing how the disclosure rules were adapted to align with the OECD Guidelines and UN Guiding Principles).

To apply the Taxonomy formula, companies also must know what information is *material* (such that reporting is required). The EU ESG framework expands on existing definitions of financial materiality in four important ways. Under the Taxonomy, companies now must focus on (i) non-financial information (e.g., information on Social or Governance issues); (ii) information beyond their direct control (e.g., facts relating to potential conflicts in areas where a company does business); (iii) information beyond the short-term; and (iv) information showing that an activity or impact *could* become material over time.

Hence, the Taxonomy now requires companies to go beyond standard financial accounting standards to assess *Double Materiality* by (i) assessing risks across the value chain that, over time, could potentially become material; and (ii) considering *both* financial and non-financial impacts up and down a company's value chain (EFRAG Draft ESRS 1—General Requirements 2022b, 11). "Impact materiality," in turn, requires companies to assess the scale, scope and remediability, and the likelihood of actual negative or positive effects on people and the environment of the company's operations, including its value chain (EFRAG Draft ESRS 1—General Requirements 2022b, 11–12). EFRAG also has defined the time horizons for disclosure (which companies may vary by specific disclosure): short-term coincides with the period used for financial statement reporting (which generally will be 12–18 months); medium-term is short-term through 5 years; and long-term is longer than 5 years (EFRAG Draft ESRS 1—General Requirements 2022b, 17). A potentially important requirement for mining in the Lithium Triangle is that companies also must disaggregate data (e.g., among different mining sites or affected communities) to make materiality disclosures on sustainability clear (EFRAG Draft ESRS 1—General Requirements 2022b, 13).

In considering ESG disclosures and strategies under the Taxonomy, companies in the lithium value chain will have to structure their disclosures to balance the environmental and social benefits of lithium extraction and use with the environmental and social risks associated with the life cycle of lithium mining.

- As to biodiversity and ecosystems, companies will be required to assess (across the short-, medium-, and long-term) and report on (i) material or potentially material positive and negative impacts; (ii) actions taken to monitor, prevent, mitigate or remediate adverse impacts, and to protect and restore biodiversity and ecosystems; (iii) corporate plans and capacity to adapt business models and operations to preserve and restore biodiversity and ecosystems (including by testing various scenarios); and (iv) the effects of biodiversity and ecosystems-related risks and opportunities on the company's performance, and its ability to create enterprise value (EFRAG Draft ESRS E4—Biodiversity and ecosystems 2022c, 8–11).
- As to social objectives regarding affected communities, companies will be required to consider the actual and potential positive and negative impacts of their activities. This will apply *both* to how communities may actually be affected

(and how those affects will be identified, monitored and, as needed, remediated), *and* the processes by which affected communities are and will be engaged as to potential positive and negative impacts. (EFRAG Draft ESRS S3—Affected Communities 2022e, 4–7).

The central importance of EU ESG framework is that it forces companies to assess and analyze the potential negative environmental and social impacts of lithium extraction *before* claiming that lithium mining is a positive activity for climate change mitigation under the Taxonomy. The calculus built into the Taxonomy creates a positive economic incentive for companies to invest in activities that *reduce* potential negative ESG impacts of lithium mining—for example, better water usage and management, better FPIC processes, and improved protection of communities throughout the entire product life cycle of lithium mining in the Lithium Triangle. Hence, the EU ESG framework also amplifies the goals of the Escazu Agreement.

10.5 Strategic ESG Approaches in the Lithium Triangle

Perhaps the most challenging aspects of the EU ESG framework are the demands relating to disclosures around long-term and potential negative impacts and events. This is new territory for many businesses, but this has long been a part of World Bank and EU risk analyses. For example, in focusing on the peace building and development challenges posed by fragile states, the World Bank has highlighted risks associated with natural resource degradation and depletion. Since 2009, in countries with World Bank development projects, almost 50% of conflict situations relate to natural resource degradation or depletion (World Bank Group 2021, viii, xii, 12–13). World Bank data also suggests that devolving resource rights to local communities can both reduce resource degradation and the risk of conflict (World Bank Group 2021, 12–13). Similarly, EU authorities have made clear that sustainability risks relating to Social Objectives can contribute to "reputational risk; markets risk; liquidity risk; credit risk; regulatory risk; operational risk; and insurance risk" (EFRAG Draft ESRS S3—Affected Communities 2022e, 4; EU Platform on Sustainable Finance 2022, 23, 24–25 (citing a Guidance Note by the German Financial Regulator (BaFin) stating that "sustainability should not be restricted to climate issues; other environmental and social trends may also present serious financial risks" (BaFin 2020, 12)).

In the context of the World Bank and other UN agencies, the risks of ecosystems and communities collapsing or falling into destructive conflict provide a basis for creating long-term peace building, conflict prevention and post-conflict prevention processes and programs. For example, data relating to societal breakdowns linked to atrocity outbreaks show how long-term data techniques can be used to assess future risks, and offer important lessons for approaching ESG strategies. Atrocity prevention models use data on governance, social, environmental, and economic risk factors to produce estimates of future risks of societal breakdowns (e.g., atrocity outbreaks, civil war, and ethnic strife)—usually assessing the risk of outbreak within 18–60 months. These models use public data sets and information from CSOs, international organizations and governments, and are reliable in shaping development and other strategies used by the World Bank, OECD and governments (see Albright and Cohen 2008, 19, 24–26; Hegre et al. 2021, 3–7). Significantly, this data confirms that *prevention* is possible in terms of using policies and processes to reduce factors that otherwise would make societal breakdown more likely. One recent study, for instance, showed that certain types of truth and reconciliation processes correlated to reductions in atrocity risk factors across a broad collection of countries (AIPG 2020, 22–27; see generally Waller 2016, 135–350 (explaining upstream, midstream and downstream prevention techniques)).

These approaches suggest that there should be ways to design long-term strategies for enhancing ESG value and reducing business risk with regard to lithium mining in Latin America. These strategies would address FPIC processes, monitoring, and impact assessment issues under both the Escazu Agreement and EU ESG framework, and could suggest best practices for addressing other critical mineral mining in other countries. *First*, Argentina, Chile and Bolivia each should create an ombudsperson with respect to mineral extraction and responsible business practices. The model for this would be the Canadian Ombudsperson for Responsible Enterprises (CORE) (Order of Council 2019). The ombudsperson would be impartial and independent, and would serve as a "national contact point" under the OECD Guidelines for Multinational Enterprises. The ombudsperson could—

- Advise companies on creating responsible business practices, including by agreeing to adhere to and implement the UN Guiding Principles on Business and Human Rights and the OECD Guidelines.
- Seek to have companies commit to best practices even where this exceeds local statutory or regulatory requirements.
- Convene CSOs, community members, companies, and government officials to address ESG issues.
- Monitor the FPIC process as to existing and new projects—including by setting minimum FPIC standards as to how disclosures should be made and how Indigenous communities should provide feedback.
- Explore ways to reduce risks of conflict as to affected communities in the Lithium Triangle.
- Monitor and examine ways to improve the CSO enabling environment, and collect and publish data relating to attacks on CSOs or civil society defenders in the Lithium Triangle.
- Address gaps in scientific and socio-economic data relating to specific sites to ensure sufficient data to support meaningful ESG analysis of the opportunities and risks associated with those sites—including as to reducing water usage and improving water management, and biodiversity and ecosystem management and protection.
- Convene parties to consider methods—even capital intensive methods—to reduce the environmental effects of lithium brine processes.

- Encourage development of data on lithium demand, lithium pricing, and effects on revenue flows to local communities and governments.
- Encourage development of data relating mining termination scenarios, including post-closure remediation costs.
- The ombudsperson also should be able to mediate disputes among affected parties. As an initial matter, the ombudsperson should address the gaps in scientific and socio-economic data relating to specific sites within the Lithium Triangle, and could even help companies understand the level of specificity that would be required to allow the FPIC process to proceed.

Second, whether through an ombudsperson or not, mining companies and CSOs should create a forum for regular meetings relating to the FPIC process and information disclosure relating to ESG issues. *Third*, companies, CSOs and representatives of affected communities should consider entering into a binding accord to address disputes arising from lithium mining operations. Precedent for this approach exists. In 2013, companies in the ready-made garment industry (manufacturers, wholesalers, and retailers) agreed to address building conditions in partnership with labor organizations representing workers in Bangladesh working in unsafe buildings, which had experienced fires and building collapses over a course of many years costing the lives of thousands of workers. The Accord addressed issues that local authorities in Bangladesh had been unable to solve.

Under that Accord, the industry agreed to ensure a safe working environment in thousands of factories tied into the value chain of the ready-made garment industry. The Accord created an independent factory inspection program focused on building and fire safety, and public disclosure relating to factories, inspection reports, and corrective actions ordered and taken. The industry committed money for building safety repairs—estimated to cover over 2300 factory sites (Senate Minority Report 2020, 26)—and agreed to use industry's relationships with factory owners to ensure that repairs were made. The industry agreed to facilitate elected worker safety committees being formed at factories and to training for workers as to their rights. The industry and unions formed a steering committee (with a neutral member provided by the International Labor Organization). Finally, the parties agreed to establish a dispute resolution process relating to worker claims arising under the Accord, which would be subject to binding arbitration—which process was funded by the industry (Bangladesh Accord 2013).

The advantage of this type of approach is that it can provide a way for companies to better understand and control their risks, including by shielding certain of those risks from local legal regimes that they may not fully trust—or for which the potential outcomes are difficult to accurately predict. This could be particularly relevant in the Lithium Triangle where the Escazu Agreement has created duties and obligations that, under certain circumstances, could give rise to claims. Those claims are difficult to estimate, however, given that the treaty is new and it is unclear what types of implementing legislation will be used in various countries.

What an ombudsperson, a forum for industry-CSO meetings, and an accord covering dispute resolution all would have in common is a long-term focus for helping companies understand how to improve the FPIC and environmental impact assessment process in the Lithium Triangle. These approaches not only would address the goals of the Escazu Agreement, but could make it more likely that the ESG benefits for lithium mining could be achieved while also increasing the likelihood that long-term ESG risks could be better managed so as to potentially generate value for shareholders.

10.6 Conclusion

The Lithium Triangle may present a unique opportunity to test how EU ESG concepts work in the field. First, the Lithium Triangle has not seen the levels of violence directed at CSO defenders in environmental and mineral development contexts that have been documented in places like Brazil, Colombia, Peru and Mexico (see generally Global Witness Archive 2016–2021). Lower levels of violence provide a good foundation for prevention planning with respect to potential conflicts. Second, Argentina, Bolivia and Chile have active CSO communities capable of reporting on FPIC and environmental issues relating to lithium extraction. Again, the presence of a CSO sector can play a significant role in conflict prevention. Third, the Escazu Agreement means that companies involved in lithium extraction in Latin America are subject to legal duties going beyond disclosure and diligence—the focus of the EU ESG framework. This creates added risks (i.e., legal and reputational), and a corresponding incentive to manage those risks over the long-term. Finally, the trailblazing EU ESG framework, currently the most detailed mandatory ESG disclosure framework, will challenge companies to provide extensive details on how the ESG benefits of lithium extraction compare to environmental and social costs along that same value chain—and to implement human rights safeguards relating to workers and communities in that value chain. These ESG-driven details then, among other things, will allow CSOs to measure corporate and lender performance in the lithium mining sector against the duties set forth in the Escazu Agreement.

These factors could be used to create strategies for managing lithium development capable of satisfying both the Escazu Agreement and EU ESG requirements. This could be particularly true in places like Bolivia, where lithium development has not ramped-up, and where the government has not settled on a hard and fast policy for lithium development. Moreover, it might be possible to use the unique opportunities for engagement in the Lithium Triangle to drive investment and financing, including by the World Bank, regional development banks, or other sovereigns and sovereign wealth funds that might be anxious to show how climate-related sustainable development can be used to attract capital while responsibly managing long-term business risks.

References

Agusdinata, Datu, Wenjuan Liu, Hallie Eakin, and Hugo Romero. 2018. Socio-environmental impacts of lithium mineral extraction: Towards a research agenda. *Environmental Research Letters* 13 (2018) No. 123001.

Albright, Madeleine, and William Cohen. 2008. *Preventing genocide: A blueprint for U.S. Policymakers*. Washington, D.C.: USHMM.

Auschwitz Institute for the Prevention of Genocide and Mass Atrocities (AIPG). 2020. *Truth commissions and their contributions to atrocity prevention*. New York: AIPG. https://www.auschwitzinstitute.org/wpcontent/uploads/2020/05/AIPG-TruthCommissionsReport.pdf.

Bangladesh Ready-Made Garment Industry Accord. 2013. https://www.industriall-union.org/sites/default/files/uploads/documents/2013-05-13_-_accord_on_fire_and_building_safety_in_bangladesh_0.pdf.

Berg, Ryan. 2021. South America's lithium triangle: Opportunities for the Biden administration. *Center for Strategic & Int'l Studies* 2021. https://www.csis.org/analysis/south-americas-lithium-triangle-opportunities-biden-administration.

Bundesanstalt fur Finanzdienstleistungsaufsicht (BaFin). 2020. *Guidance notice on dealing with sustainability risks*. Berlin: BaFin. https://www.bafin.de/SharedDocs/Veroeffentlichungen/EN/Meldung/2019/meldung_191220_MB_Nachhaltigkeitsrisiken_en.html.

Canadian Order of Council. 2019. *Mandate for the ombudsperson for responsible Enterprise*. https://orders-in-council.canada.ca/attachment.php?attach=38652&lang=en.

Civicus. 2020. Argentina's former government accused of spying on journalists, civil society leaders and others. https://monitor.civicus.org/updates/2020/09/14/argentinas-former-government-accused-spying-journalists-civil-society-leaders-and-others/.

———. 2022. *Bolivia: Intimidation of environmental defenders and journalists*. https://monitor.civicus.org/updates/2022/06/27/bolivia-intimidation-environmental-defenders-and-journalists/.

Convention on Access to Information, Public Participation in Decision-Making and Access to Justice in Environmental Matters. 1998. 25 June 1998 at Aarhus, Denmark, entry into force 30 October 2001. 2161 U.N. Treaty Series 447, 1998.TREATIES-XXVII.

Escosteguy, Melisa, Araceli Clavijo, Water Paz and Marc Hufty. 2022. "We are not allowed to speak": Some thoughts about a consultation process around lithium mining in Northern Argentina. 11 *The Extractive Industries & Society* (2022). No. 101134. https://www.sciencedirect.com/science/article/abs/pii/S2214790X22000983.

European Commission. 2020. Inception Impact Assessment, Sustainable Corporate Governance. Ref. Ares 4034032-30/07/2020.

European Financial Reporting Advisory Group (EFRAG). 2021. *Final report: Proposals for a relevant and dynamic EU sustainability reporting standard-setting*. https://www.efrag.org/Assets/Download?assetUrl=%2Fsites%2Fwebpublishing%2FSiteAssets%2FEFRAG%2520PTF-NFRS_MAIN_REPORT.pdf.

———. 2022a. Draft European sustainability reporting standards, cover letter. Submitted to European Commission November 22, 2022. https://www.efrag.org/Assets/Download?assetUrl=%2Fsites%2Fwebpublishing%2FSiteAssets%2F01%2520EFRAG%2527s%2520Cover%2520Letter%2520to%2520the%2520first%2520set%2520of%2520ESRS%252022%2520November%25202022.pdf.

———. 2022b. Draft European sustainability reporting standards, ESRS 1—General requirements. Submitted to European Commission November 22, 2022. https://www.efrag.org/Assets/Download?assetUrl=%2Fsites%2Fwebpublishing%2FSiteAssets%2F06%2520Draft%2520ESRS%25201%2520General%2520requirements%2520November%25202022.pdf.

———. 2022c. Draft European sustainability reporting standards, ESRS E4—Biodiversity and ecosystems. Submitted to European Commission November 22, 2022. https://www.efrag.org/Assets/Download?assetUrl=%2Fsites%2Fwebpublishing%2FSiteAssets%2F11%2520Draft%2520ESRS%2520E4%2520Biodiversity%2520and%2520ecosystems%2520November%25202022.pdf.

———. 2022d. Draft European sustainability reporting standards, ESRS G1—Business conduct. Submitted to European Commission November 22, 2022. https://www.efrag.org/Assets/Download?assetUrl=%2Fsites%2Fwebpublishing%2FSiteAssets%2F17%2520Draft%2520ESRS%2520G1%2520Business%2520Conduct%2520November%25202022.pdf.

———. 2022e. Draft European sustainability reporting standards, ESRS S3—Affected communities. Submitted to European Commission November 22, 2022. https://www.efrag.org/Assets/Download?assetUrl=%2Fsites%2Fwebpublishing%2FSiteAssets%2F15%2520Draft%2520ESRS%2520S3%2520Affected%2520communities%2520November%25202022.pdf.

———. 2022f. Draft European sustainability reporting standards, explanatory note. Submitted to European Commission November 22, 2022. https://www.efrag.org/Assets/Download?assetUrl=%2Fsites%2Fwebpublishing%2FSiteAssets%2F03%2520Explanatory%2520note%2520Fist%2520set%2520of%2520ESRS%2520Article%252029%2520b_last.pdf.

European Parliament. 2014. *Directive as regards disclosure of non-financial and diversity information by certain large undertakings and groups*. Directive 2014/95/EU.

———. 2019. *Regulation on sustainability-related disclosures in the financial services sector*. Regulation 2019/2088.

———. 2020. *Regulation on the establishment of a framework to facilitate sustainable investment*. Regulation 2020/852.

———. 2021. *Proposed directive as regards corporate sustainability reporting*. 2021/0104 COD.

———. 2022. *Proposed directive on corporate sustainability due diligence*. 2022/051 COD.

European Parliament Press Release. 2022. *Sustainable economy: Parliament adopts new reporting rules for multinationals*. https://www.europarl.europa.eu/news/en/press-room/20221107IPR49611/sustainable-economy-parliament-adopts-new-reporting-rules-for-multinationals.

European Union Platform on Sustainable Finance. 2022. *Final report by subgroup 4: Social taxonomy*. https://ec.europa.eu/info/sites/default/files/business_economy_euro/banking_and_finance/documents/280222-sustainable-finance-platform-finance-report-social-taxonomy.pdf.

Global Witness. 2016–2021. *Annual report archive*. https://www.globalwitness.org/en/campaigns/environmental-activists/land-and-environmental-defenders-annual-report-archive/.

———. 2021. *Last line of defence: The industries causing the climate crisis and attacks against land and environmental defenders*. https://www.globalwitness.org/en/campaigns/environmental-activists/last-line-defence/.

Guterres, Antonio. 2021. *U.N. Secretary General Statement on entry into force of Escazu Agreement*. https://www.un.org/sg/en/content/sg/statement/2021-04-22/secretary-generals-message-marking-the-entry-force-of-the-escaz%C3%BA-agreement.

Hegre, Havard, Havard Nygard, and Peder Landsverk. 2021. Can we predict armed conflict? How the first 9 years of published forecasts stand up to reality. *Int'l Studies Quarterly* 2021: 1–9.

Inter-American Court of Human Rights. 2017. *The Environment and Human Rights (State Obligations in Relation to the Environment, in the Context of the Protection and Guarantee of the Rights to Life and to Personal Integrity)*, Advisory Op. OC-23/18, Inter-Am. Ct. H.R. (ser. A) No. 23.

International Finance Corporation. 2012. The business case for sustainability.

Kelly, Jarrod, Michael Wang, Qiang Dai, and Olumide Winjobi. 2021. Energy, greenhouse gas, and water life cycle analysis of lithium carbonate and lithium hydroxide monohydrate from brine and ore resources and their use in lithium ion batter cathodes and lithium ion batteries. Argonne Nat'l Laboratory. *Journal of Resources, Conservation & Recycling* 174 (2021) No. 105762. https://www.sciencedirect.com/science/article/pii/S0921344921003712?via%3Dihub.

Lopez-Calva, Luis Felipe. 2022. *Lithium in Latin America: A new quest for "El Dorado"?* https://www.undp.org/latin-america/blog/graph-for-thought/lithium-latin-america-new-quest-el-dorado.

Molen, Isabel. 2022. *Bolivia: Pursuing sustainable lithium mining, human rights pulse*. https://www.csis.org/blogs/development-dispatches/bolivia-pursuing-sustainable-lithium-mining.

Marchegiani, Pia, Jasmin Hoglund Hellgren, and Leandro Gomez. 2019. *Lithium extraction in Argentina: A case study on the social and environmental impacts*. Buenos Aires: Fundacion

Ambiente Y Recursos Naturales. https://farn.org.ar/wp-content/uploads/2019/05/DOC_LITHIUM_ENGLISH-1.pdf.

Organization of American States. 1988. Additional protocol to the American Convention on Human Rights in the area of economic, social and cultural rights (Protocol of San Salvador). Nov. 14, 1988 San Salvador, El Salvador. 28 I.L.M. 156 (1989).

Organization for Economic Co-operation and Development (OECD). 2011. *Guidelines for multinational enterprises*. https://www.oecd.org/daf/inv/mne/48004323.pdf.

Purdy, Caitlin, Rodrigo Castillo. 2022. The future of Mining in Latin America: Critical minerals and the global energy transition. Brookings Institution: LTRC Project. https://www.brookings.edu/research/the-future-of-mining-in-latin-america-critical-minerals-and-the-global-energy-transition/.

Regional Agreement on Access to Information, Public Participation and Justice in Environmental Matters in Latin America and the Caribbean. 2018. 4 March 2018 at Escazu, Costa Rica, entered into force, 22 April 2020. UN Treaty Index C.N.195.2018.TREATIES-XXVII.18. The state parties are Antigua and Barbuda, Argentina, Bahamas, Barbados, Belize, Bolivia, Brazil, Chile, Colombia, Costa Rica, Cuba, Dominica, Dominican Republic, Ecuador, El Salvador, Grenada, Guatemala, Guyana, Haiti, Honduras, Jamaica, Mexico, Nicaragua, Panama, Paraguay, Peru, Saint Kitts and Nevis, Saint Lucia, Saint Vincent and the Grenadines, Suriname, Trinidad and Tobago, Uruguay and Venezuela. The treaty took effect upon the eleventh state party ratifying the agreement. Escazu Agreement 2018, Arts. 21–22 & Annex 1.

Riofrancos, Thea. 2022. *Chile: White gold rush*. https://www.chathamhouse.org/publications/the-world-today/2022-06/chile-white-gold-rush.

Squire, Carter, Kelsey Landau, and Robin Lewis. 2020. *Uncommon ground: The impact of natural resource corruption on indigenous peoples*. Brookings Institution. https://www.brookings.edu/blog/up-front/2020/08/07/uncommon-ground-the-impact-of-natural-resource-corruption-on-indigenous-peoples/.

Task Force on Justice. 2019. *Justice for all*. Task Force on Justice: Final Report. https://www.justice.sdg16.plus/report-old2022.

United Nations Economic Commission for Latin America and the Caribbean (ECLAC). 2022. *Implementation guide for regional agreement on access to information, public participation and justice in environmental matters in Latin America and the Caribbean*. LC/TS.2021/221, No. 21-00641.

———. 2015. *Preliminary document of the regional instrument on access to information, participation and justice on environmental matters in Latin America and the Caribbean*. LC/L.3987, No. 15-00260.

United Nations General Assembly. 1999. Declaration on the right and responsibility of individuals, groups and organs of society to promote and protect universally recognized human rights and fundamental freedoms (UN resolution on human rights defenders). A/RES/53/144.

———. 1992. Report of the United Nations conference on environment and development, annex I: Rio declaration on environment and development. A/CONF.151/26 (Vol. I).

———. 2015. 2030 agenda for sustainable development. A/RES/70/1. https://www.un.org/sustainabledevelopment/peace-justice/.

———. 2007. Universal declaration on the rights of indigenous peoples. A/RES/61/295.

United Nations Office of the High Commissioner on Human Rights (UN OHCHR). 2011. *UN guiding principles on business and human rights*. https://www.ohchr.org/sites/default/files/documents/publications/guidingprinciplesbusinesshr_en.pdf.

United States Geological Survey (USGS). 2022. Mineral commodity summary. https://pubs.usgs.gov/periodicals/mcs2022/mcs2022-lithium.pdf.

United States Senate. Committee on Foreign Relations, Minority Staff Report. 2020. *Seven years after Rana Plaza, significant challenges Remain*, 116th Cong., 2d Sess. The Minority Report can be found at https://www.govinfo.gov/content/pkg/CPRT-116SPRT39906/pdf/CPRT-116SPRT39906.pdf

Waller, James. 2016. *Confronting evil: Engaging our responsibility to prevent genocide*. Oxford: Oxford University.

Walter, Mariana and Lucrecia Wagner. 2021. Mining Struggles in Argentina: The keys to a successful story of mobilization. 8 *The Extractive Industries & Society*. No. 100940. https://www.sciencedirect.com/science/article/pii/S2214790X21001027.

World Bank Group. 2020. *Minerals for climate action: The mineral intensity of the clean energy transition*. https://www.worldbank.org/en/topic/extractiveindustries/brief/climate-smart-mining-minerals-for-climate-action.

———. 2021. *The natural resource degradation and vulnerability nexus: An evaluation of the World Bank's support for sustainable and inclusive natural resource management (2009–19)*. https://ieg.worldbankgroup.org/sites/default/files/Data/Evaluation/files/NaturalResourceDegradationVulnerabilityNexus.pdf.

Owen Pell is President of the Board of the Auschwitz Institute for the Prevention of Genocide and Mass Atrocities (www.auschwitzinstitute.org) and a member of the Board of the International Peace Institute (www.ipinst.org). He is a retired partner of White & Case LLP, where he represented governments, banks, and companies in complex, cross-border disputes and handled or advised on major cases involving business and human rights, including as to World War II and the Holocaust, Apartheid-era South Africa, and African slavery in the United States. He has lectured widely on ESG and business and human rights issues. He also formulated a proposal that was endorsed by the European Parliament for addressing claims to works of art looted from individuals during the Holocaust.

Index

S. Kalantzakos (ed.), *Critical Minerals, the Climate Crisis and the Tech Imperium*, Archimedes 65, https://doi.org/10.1007/978-3-031-25577-9

D

E

The manufacturer's authorised representative in the EU is Springer Nature Customer Service Centre GmbH, Europaplatz 3, 69115 Heidelberg, Germany. If you have any concerns regarding our products, please contact ProductSafety@springernature.com

Printed and bound by CPI Group (UK) Ltd, Croydon, CR0 4YY
28/09/2026
02224075-0004